Functional Properties
of Proteins and Lipids

ACS SYMPOSIUM SERIES **708**

Functional Properties of Proteins and Lipids

John R. Whitaker, EDITOR
University of California at Davis

Fereidoon Shahidi, EDITOR
Memorial University of Newfoundland

Agustin López Munguia, EDITOR
Universidad Nacional Autonoma de Mexico

Rickey Y. Yada, EDITOR
University of Guelph

Glenn Fuller, EDITOR
U.S. Department of Agriculture

Developed from a symposium sponsored by the Division
of Agricultural and Food Chemistry
of the American Chemical Society
at the Fifth Chemical Congress of North America
Cancun, Quintana Roo, Mexico
November 11–15, 1997

American Chemical Society, Washington, DC

Library of Congress Cataloging-in-Publication Data

Functional properties of proteins / John R. Whitaker ... [et al.], editors.

   p.   cm.—(ACS symposium series, ISSN 0097-6156; 708)

   "Developed from a symposium sponsored by the Division of Agricultural and Food Chemistry of the American Chemical Society at the Fifth Congress of North America, Cancun, Quintana Roo, Mexico, November 11–15, 1997."

   Includes bibliographical references and index.

   ISBN 0–8412–3584–8 (alk. paper)

   1. Proteins.

   I. Whitaker, John R.,   II. Chemical Congress of North America (5th : 1997 : Cancun, Mexico)   III. Series.

QD431.F86   1998
547'.75—dc21   98–25964
                                                                          CIP

The paper used in this publication meets the minimum requirements of American National Standard for Information Sciences—Permanence of Paper for Printed Library Materials, ANSI Z39.48–1984.

PRINTED IN THE UNITED STATES OF AMERICA

QD431
F86
1998
CHEM

Advisory Board

ACS Symposium Series

Foreword

THE ACS SYMPOSIUM SERIES was first published in 1974 to provide a mechanism for publishing symposia quickly in book form. The purpose of the series is to publish timely, comprehensive books developed from ACS sponsored symposia based on current scientific research. Occasionally, books are developed from symposia sponsored by other organizations when the topic is of keen interest to the chemistry audience.

Before agreeing to publish a book, the proposed table of contents is reviewed for appropriate and comprehensive coverage and for interest to the audience. Some papers may be excluded in order to better focus the book; others may be added to provide comprehensiveness. When appropriate, overview or introductory chapters are added. Drafts of chapters are peer-reviewed prior to final acceptance or rejection, and manuscripts are prepared in camera-ready format.

As a rule, only original research papers and original review papers are included in the volumes. Verbatim reproductions of previously published papers are not accepted.

ACS BOOKS DEPARTMENT

Contents

FAT AND OIL FUNCTIONALITY;
PHYSIOLOGICAL FUNCTIONALITY

INDEXES

Preface

It is often unclear when a new area of science and/or technology begins. But it is reasonably clear that the study of food functionality is of recent origin, beginning about the mid 1960s. This is about the time that biological functions of proteins as enzymes, hormones, transmitters, etc. began to be explored. However, it was in the 1980s before biochemistry textbooks began to divide proteins for discussion into their biological functional contributions. In 1979, Pour-El (1) edited an ACS Symposium Series volume entitled *Functional and Protein Structure* in which he and the 12 chapter authors began to lay the foundation for future directions and advances in this field. In 1976, Pour-El (2) defined functionality as "any property of a substance, besides its nutritional ones, that affects its utilization." Although not inherent in the definition, much of the subsequent research on food functionality was focused on proteins, leaving aside the contributions of lipids and carbohydrates. Some might question also whether the nutritional properties should not be included in the definition.

From the beginning of food functionality studies separate pathways, including instrumentation and methodology, developed between those investigating functionality of food systems and those investigating functionality of isolated component systems. Both approaches, the empirical and the model, have contributed to a better understanding of the factors that contribute to the functional quality of foods. It is clear that we must first understand the chemical and physical properties of the principal functional components of foods in order to maximize their utilization in foods. In the case of proteins, for example, it has been shown repeatedly that the amino acid sequence, and post-translational modification of that sequence, markedly affect the food functionality properties, just as the biological properties are affected. In 1991, Kato (3) demonstrated the significance of macromolecular interaction and protein stability on functional properties of food proteins, raising the possibility that selected proteins could be tailored genetically to provide required functionality. The same has recently been demonstrated for lipids, either by genetic engineering or by selective crystallization (4). A recent publication edited by Parris et al. (5) provides numerous examples for the importance of this approach, including chemical, enzymatic and genetic modifications of proteins (6–10). We hope that the current book will add to this knowledge.

Some other areas of food functionality have not advanced quite so rapidly as those above. These include common methods and conditions for determining functionality so that results can be compared among researchers; general understanding of fundamental principles that determine food functionality not only by scientists and teachers but also those who apply the technology either through ingredient manufacture or in consumable products. Kolar et al. (*11*) have shown that commercial spray-dried soy protein isolates can differ markedly in functional properties, something most users of the products do not know about.

This book brings the subject of food functionality up to date, based on current research of leading scientists in Canada, Mexico, and the United States, and their associates around the world. Much of the research reported is that of the individual authors and is primary data.

The book is divided into four sections. The first section—four chapters on fundamental properties and instrumentation methods—deals with targeted research and predictions of the relationships of molecular properties of food ingredients to their functional properties and behavior in foods. The second section—three chapters, plant protein functionalities—discusses production of high-protein flours for use as milk substitutes and the functional properties of soy protein, and the effect of acylation on modifying the functional properties of fax proteins.

The third section—seven chapters, animal protein functionalities—describes the structure–functional relationships of milk and whey proteins and of meat proteins. All the chapters relate fundamental properties of the proteins to their use in foods or as coatings for foods. Chapter 13 addresses the industrial uses of whey protein concentrates and isolates, noting numerous factors that affect their functionalities in foods.

The fourth section—three chapters, fat and oil functionality; physiological functionality—is not found in most books on functionality. The mouthfeel of fats and oils, and their incorporation of flavor constituents are very important in food selection, as for example, the "marbling" of a prime rib steak over that of a less quality one. The last chapter on physiological functionality of food components should be valuable for all readers of this book, including nutritionists.

This book, because of its unique blend of fundamental and applied aspects of food ingredient functionality, should be attractive to a wide range of food scientists, food technologists, and industrial persons.

Literature Cited

1. *Functionality and Protein Structure;* Pour-El, A., Ed.; ACS Symposium Series 92; American Chemistry Society: Washington, DC, 1979; 243 pp.
2. Pour-El, A. In *World Soybean Research;* Hill, L. D., Ed.; Interstate: 1976; pp 918–948.
3. Kato, A. In *Interactions of Food Proteins;* Parris N.; Barford, R., Eds.; ACS Symposium Series 454; American Chemical Society: Washington, DC, 1991; pp 13–24.
4. German, J. B.; Dillard, C. J. *Food Technol.* **1998,** *52(2),* 33–34, 36–38.
5. *Macromolecular Interactions in Food Technology;* Parris, N.; Kato, A.; Creamer, L. K.; Pearce, J., Eds.; ACS Symposium Series 650; American Chemical Society: Washington, DC, 1996; 304 pp.

6. Vojdani, F.; Whitaker, J. R. In *Macromolecular Interactions in Food Technology;* Parris, N.; Kato, A.; Creamer, L. K.; Pearce, J., Eds.; ACS Symposium Series 650; American Chemical Society: Washington, DC, 1996; pp 210–229.

7. Aoki, T. In *Macromolecular Interactions in Food Technology;* Parris, N.; Kato, A., Creamer, L. K.; Pearce, J., Eds.; ACS Symposium Series 650; American Chemical Society: Washington, DC, 1996; pp 230–242.

8. Kato, A.; Nakamura, S.; Takasaki, H.; Maki, S. In *Macromolecular Interactions in Food Technology;* Parris, N.; Kato, A.; Creamer, L. K.; Pearce, J., Eds.; ACS Symposium Series 650; American Chemical Society: Washington DC, 1996; pp 243–256.

9. Sequro, K.; Nio, N.; Motoki, M. In *Macromolecular Interactions in Food Technology;* Parris, N.; Kato, A.; Creamer, L. K.; Pearce, J., Eds.; ACS Symposium Series 650; American Chemical Society: Washington, DC, 1996; pp 271–280.

10. Hill, J. P.; Boland, M. J.; Creamer, L. K.; Anema, S. G.; Otter, D. E.; Paterson, G. R.; Lowe, R.; Motion, R. L.; Thresher, W. C. In *Macromolecular Interactions in Food Technology;* Parris, N.; Kato, A.; Creamer, L. K.; Pearce, J., Eds.; ACS Symposium Series 650; American Chemical Society: Washington, DC, 1996; pp 281–294.

11. Kolar, C. W.; Richert, S. H.; Decker, C. D.; Steinke, F. H.; Vander Zanden, R. J. In *New Protein Foods; Vol. 5. Seed Storage Proteins;* Altschul, A. M.; Wilcke, H. L., Eds.; Academic: Orlando, FL, 1985; pp 259–299.

JOHN R. WHITAKER
Department of Food Science and Technology
University of California
Davis, CA 95616

FEREIDOON SHAHIDI
Department of Biochemistry
Memorial University of Newfoundland
St. John's, Newfoundland A1B 3X9, Canada

AGUSTIN LOPEZ MUNGUIA
Instituto de Biotechnologia UNAM
Cuernavaca, Morelos 62250, Mexico

RICKEY Y. YADA
Department of Food Science
University of Guelph
Guelph, Ontario N1G 2W1, Canada

GLENN FULLER
U.S. Department of Agriculture/ARS
Western Regional Research Center
Albany, CA 94710

FUNDAMENTAL PROPERTIES
AND INSTRUMENTAL METHODS

Chapter 1

Molecular Bases of Surface Activity of Proteins

S. Damodaran and L. Razumovsky

Department of Food Science, University of Wisconsin, Madison, WI 53706

The first critical event in the creation of protein-stabilized emulsions and foams is adsorption of proteins at the oil-water and air-water interfaces. This initial event is dependent on the molecular flexibility of proteins. The dynamic equation-of-state, Θ, of proteins at the air-water and oil-water interfaces is directly related to flexibility. The Θ value of proteins, which represents dynamic reduction in interfacial free energy per mg of adsorbed protein (ergs/mg), at the air-water interface is also linearly correlated with the emulsifying properties, foaming properties, proteolytic digestibility, and the adiabatic compressibility of proteins. Based on these correlations, it is proposed that the Θ values of proteins can be used as a predictor of several functional properties of proteins.

A majority of foods can be categorized as emulsions and foams. These are two-phase systems in which one of the phases (oil or air) is dispersed in the other phase. The stability of these dispersed systems depends on the presence of a surfactant which can significantly reduce the interfacial tension between the phases. Two types of surfactants (emulsifiers or foaming agents) are being used in foods. These are low molecular weight surfactants, such as phospholipids, mono- and diglycerides, sorbitan monostearate, etc., and polymeric surfactants, such as proteins and certain gums. Although low molecular weight surfactants are more effective than proteins in reducing the interfacial tension, foams and emulsions created by them are not as stable as those formed by proteins. This is because proteins, in addition to lowering interfacial tension, can form a membrane-like viscoelastic film around oil droplets or air bubbles via noncovalent interactions and disulfide cross-linking, which is not possible in the case of low molecular weight surfactants.

The surface activity of proteins emanates from the fact that they are amphiphathic. That is, they contain both hydrophilic and hydrophobic groups/residues, which can orient themselves at interfaces. However, a cursory examination of the amino acid compositions of various proteins indicates that the percent distribution of polar and nonpolar amino acid residues is very similar in various proteins. Therefore, large differences in the surface activities of various proteins must be related to differences in physicochemical properties, such as molecular flexibility, conformational stability at interfaces, and the nature of distribution of hydrophilic and hydrophobic residues on a protein's surface. The physicochemical properties of a protein's surface and the molecular flexibility of a protein in turn are dependent on the way the polypeptide chain is folded in the three-dimensional space, i.e., its conformation. If this is the case, intuitively then the surface activities of poorly functional proteins, such as plant proteins, may be improved by intelligently altering their conformation using thermal, enzymatic, and genetic modification approaches. However, to apply these approaches successfully, it is essential first of all to know how much hydrophobic or hydrophilic surface a protein ought to have or how flexible certain regions of a protein must be in order for it to exhibit improved functionality. These relationships are not very well understood.

Protein Adsorption at Interfaces

The first step in the formation of protein-stabilized foams and emulsions is adsorption of proteins at air-water and oil-water interfaces. Generally, when a protein solution is exposed to a gas (air) or oil phase, the equilibrium concentration of protein in the interfacial region is always in excess of that in the bulk aqueous phase, suggesting that protein molecules migrate spontaneously from the bulk phase to these interfaces. The rate and extent of accumulation of proteins in the interfacial region depends on several factors. First, when a protein approaches an interface, whether or not it would adsorb to the interface depends on the probability of success of each collision with the interface leading to adsorption. This is a function of the hydrophobicity /hydrophilicity ratio of the protein's surface. However, analysis of crystallographic structures of several globular proteins shows that on an average about 50% of a protein's solvent accessible surface is occupied by nonpolar amino acid residues (1). This suggests that it is not the hydrophilicity/hydrophobicity ratio of and their distribution on the protein's surface, but rather the distribution pattern of these nonpolar residues on the surface which influences adsorption of proteins. That is, single nonpolar residues distributed randomly on the protein surface may not have sufficient interaction energy to anchor the protein at the interface. If the nonpolar residues on the protein surface are seggregated in the form of patches consisting of several nonpolar residues, then collision of these patches with the air-water or oil-water interface may lead to adsorption. Thus, the initial probability of adsorption of a protein as it approaches an interface may fundamentally depend on the number, size, and distribution of nonpolar patches on the protein.

Examples of this phenomenon are shown in Figure 1, which shows the time-dependent accumulation of four structurally different proteins, *viz.* β-casein, bovine serum albumin (BSA), lysozyme, and phosvitin at the air-water interface (2). In these experiments, the concentration of the protein at the air-water interface was monitored using a surface radiotracer method and the surface pressure was recorded using the Wilhelmy plate method (3). Both these measurements were made simultaneously during adsorption of the proteins from a dilute bulk phase (1.5 µg/ml in 10 mM phosphate buffered saline solution, pH 7.0 and $I = 0.1$) to the air-water interface. Among these proteins, ß-casein adsorbed very rapidly at the air-water interface and formed a saturated monolayer with a surface excess of about 2 mg/m^2. BSA also adsorbed at a reasonably rapid rate, but formed a saturated monolayer of only about 1 mg/m^2. In the case of lysozyme, it exhibited a long lag phase before adsorbing to the interface, and formed a saturated monolayer of about 0.75 mg/m^2. On the other hand, phosvitin, a highly negatively charged random-coiled phosphoglycoprotein, did not adsorb to the air-water interface at pH 7.0. The differences in adsorption behaviors of these four proteins are related to differences in their conformation and also their surface hydrophobicity. The differences in surface hydrophobicity of these proteins impacted on the probability of each collision of the proteins with the interface leading to successful adsorption. In this regard, both β-casein and BSA are hydrophobic proteins and contain several hydrophobic patches on their surface, whereas although about 40% of lysozyme's surface is hydrophobic, it contains very few hydrophobic patches. Being highly polyanionic and completely random coil at pH 7.0, phosvitin's surface contains no hydrophobic patches and hence it did not bind to the interface.

The most critical requirement for the creation of a foam or emulsion during whipping or homogenization is the ability of a protein to rapidly decrease the free energy (tension) of the newly formed interface. Even if a protein adsorbs rapidly and accumulates at the interface, it may not necessarily be able to rapidly reduce the interfacial tension. Because, the ability of a protein to reduce the interfacial tension is not related to its concentration alone at the interface, but to its ability to unfold, undergo conformational rearrangement, and form a cohesive film via intermolecular interactions at the interface.

The time-dependent increase of surface pressure of β-casein, BSA, and lysozyme is shown in Figure 2. Among these proteins, the surface pressure of β-casein reaches a steady-state value when the surface concentration reaches its steady-state value. However, this is not the case for BSA and lysozyme. The surface pressure of both these proteins increases even after their surface concentrations have reached a steady-state value. This suggests that both BSA and lysozyme undergo slow conformational changes at the interface and such conformational changes cause additional decrease of the free energy of the interface.

The surface pressure of a protein film at the air-water interface at any given surface excess is a sum of contributions from three forces (4)

$$\Pi_{a/w} = \Pi_{kin} + \Pi_{ele} + \Pi_{coh} \qquad [1]$$

where Π_{kin}, Π_{ele}, and Π_{coh} are the contributions's from kinetic, electrostatic, and cohesive forces, respectively. Since proteins are large and therefore kinetic motions of the molecules at interfaces is slow, the contribution of Π_{kin} to surface pressure is usually very low. Above 0.1 ionic strength, the contribution of Π_{ele} is also very low because of charge neutralization. Thus, in general, Π_{coh}, i.e., the cohesive interaction between protein molecules at the interface, is the single most important contributor to surface pressure.

Surface Pressure - Surface Concentration Relationship. The onset of intermolecular interactions between adsorbed protein molecules is related to their concentration at the surface. The critical surface concentration above which cohesive interactions occur can be determined from a plot of surface concentration (Γ) against surface pressure (Π). For example, the Γ-Π plots for β-casein, BSA and lysozyme are shown in Figure 3. For these proteins, the surface concentration at which the surface pressure starts to increase is in the range of 0.4-0.5 mg/m². Below this critical concentration, where no measurable

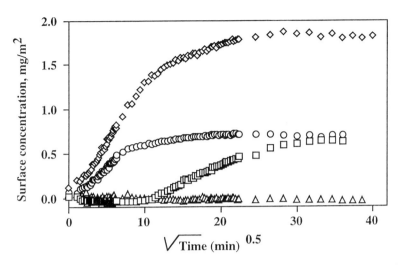

Figure 1. Time-dependent increase of surface concentration during adsorption of β-casein (◇), BSA (○), lysozyme (□), and phosvitin (△) at the air-water interface from a 1.5 μg/mL protein in 10 mM phosphate buffered saline solution at pH 7.0 and $I = 0.1$. (From ref. 2).

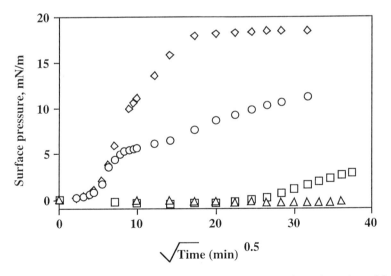

Figure 2. Time-dependent increase of surface pressure during adsorption of β-casein (◇), BSA (○), lysozyme (□), and phosvitin (△) at the air-water interface from a 1.5 μg/mL protein in 10 mM phosphate buffered saline solution at pH 7.0 and $I = 0.1$. (From ref. 2).

surface pressure is detected, the protein molecules at the interface exist as individual molecules and are probably in a gaseous state. The average distance between the protein molecules is so large that no intermolecular cohesive interactions between the molecules is possible and hence there is no protein film formation below this critical concentration.

Above the critical concentration, the contribution of Π_{coh} to the total surface pressure seems to be protein specific. This is reflected in the slopes of the Γ-Π curves. For example, the slope is about 14.3, 23.0, and 37.4 mN.m/mg, respectively, for β-casein, lysozyme, and BSA. The slope, which is in units of energy per mass, reflects the reduction in the free energy of the *interface* (not that of the protein) caused by adsorption of one milligram of protein under dynamic adsorption conditions. Thus, differences in the slopes apparently reflects differences in the magnitude of cohesive interactions among protein molecules as a function of protein concentration in the film. In this regard, the intermolecular cohesive interactions follows the order BSA > lysozyme > β-casein.

It is generally assumed that, being a hydrophobic and random-coil-like protein, β-casein is more surface active than the other globular proteins. This is partly true, because among all proteins studied so far, β-casein exhibits the greatest tendency to accumulate at both the air-water and oil-water interfaces. For instance, whereas the saturated monolayer coverage for β-casein at the air-water interface is about 2 mg/m^2, it is only about 1 mg/m^2 for other globular proteins. However, the data shown in Figure 3 clearly indicate that, on a per-mass-basis, BSA is more surface active than β-casein in terms of decreasing the free energy of the interface. This is fundamentally related to intermolecular cohesive interactions between protein molecules at the interface. Previously, it has been shown that the stability of protein foams is affected by the rheological properties, which is related to cohesive interactions, of protein films (5). The shear viscosity of β-casein film at the air-water interface was less than 1.0 mNs/m, whereas that of BSA was about 400 mNs/m (5). Because of its high shear viscosity, foams of lysozyme are more stable than those of β-casein.

Intuitively, the slope of the Γ-Π curves, which we define as Θ, which represents decrease in the free energy of the interface per mg of protein adsorbed, must be also related to molecular flexibility of globular proteins. The reasoning for this is as follows. If a globular protein is highly flexible, then its tendency to unfold at the interface also will be high. Upon unfolding, the newly exposed functional groups, such as nonpolar and hydrogen bonding groups, will promote cohesive interactions between the adsorbed proteins, leading to formation of a cohesive film. Thus, the extent of cohesive interactions in the film and the consequent reduction in interfacial free energy as a function of surface concentration may be related to molecular flexibility. If this argument is indeed valid, then, since β-casein is more flexible than BSA and lysozyme, the Θ value of β-casein should be greater than those of BSA and lysozyme. However, this is not the case. This anomaly may be explained as follows: β-Casein is not a typical globular protein. Although it is highly hydrophobic and flexible, intermolecular cohesive interactions among β-casein molecules at the air-water interface is low because of its tendency to undergo rapid changes in configuration, i.e., from trains to loops, at interfaces. One of the reasons for this might be its high proline content (~17%). While this kind of rapid trans-configurational changes at the interface may facilitate greater amount of its packing in the saturated monolayer than other globular proteins (about 2 mg/m² for β-casein versus about 1.0 mg/m² for globular proteins), it adversely impacts the formation of stable cohesive linkages within the β-casein film and thereby its effectiveness in reducing the surface/interface energy per mg of protein adsorbed. In other words, even though β-casein is highly flexible, it can not form a stable cohesive bonding between segments because of rapid transconfigurational changes. For truly globular proteins, however, this may not be the case. Even in a partially unfolded state at the interface, globular proteins retain a fair degree of structural rigidity. They assume a molten-globule state with substantial amount of α-helix and β-sheet structures, which do not permit rapid train-to-loop trans-configurational changes in the polypeptide chain. The lack of rapid trains-to-loops configurational changes in globular proteins may facilitate formation of stable cohesive interactions between the segments of the partially unfolded protein and this may lead to effective reduction of surface/interface free energy.

From the above arguments, it can be surmised that in order for a globular protein to form a cohesive film via intermolecular interactions, it should be flexible. The greater the flexibility of a globular protein, the higher would be the Θ value. The problem encountered with establishing a quantitative relationship between flexibility and Θ is that the molecular flexibility is not a readily quantifiable parameter. However, because flexibility manifests itself in dynamic fluctuations in volume (6), it can be indirectly assessed from compressibility of globular proteins. The partial specific adiabatic compressibility, β_s, of proteins is defined as

$$\beta_s = (\beta_o / \upsilon^\circ) \lim_{c \to 0} [(\beta/\beta_o - V_o)/c] \qquad [2]$$

where $V_o = (d-c)/d_o$, β and β_o are the adiabatic compressibility of the solution and solvent, respectively; d is the density of the solution; d_o is the density of the solvent; c is the concentration of the protein in grams per milliliter of solution; V_o is the apparent volume fraction of the solvent in solution, and υ° is the partial specific volume of the solute. The values of β and β_o can be experimentally determined from sound velocity using the relation

$$\beta = 1/du^2 \qquad [3]$$

where u is sound velocity. Knowing the values of β, β_o, and V_o, the limit function in equation 2 can be determined from the intercept of a plot of c versus $(\beta/\beta_o - V_o)/c$.

Correlation Between β_s and Θ. Gekko and Hasegawa (6) reported β_s values for several proteins, including β-casein. These values are given along with the Θ values for some of the proteins in **Table I**. The correlation between adiabatic compressibility and Θ is shown in Figure 4. The Θ of globular proteins increases linearly with increasing β_s, indicating that the surface activity of proteins is indeed highly correlative with compressibility of proteins. The least-squares linear regression of the Θ-β_s plot for globular proteins yields the following relation with a correlation coefficient of 0.86,

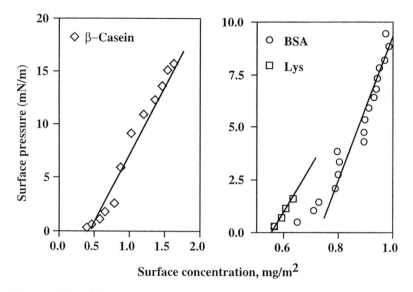

Figure 3. The surface concentration *versus* surface pressure relationships for β-casein, BSA, and lysozyme under dynamic adsorption conditions from a bulk solution to the air-water interface. The plots are derived from data shown in Figures 1 and 2.

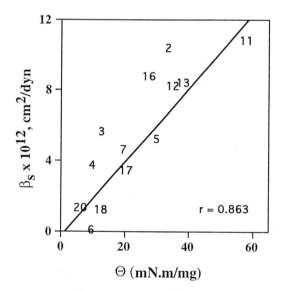

Figure 4. Plots of adiabatic compressibility, β_s, against Θ values of proteins. The numbers of the points refer to the identity of proteins in Table I. The solid line represents the least-squares linear regression for equation 4.

Table I. Compressibility and Θ values of various proteins

	Protein	Adiabatic Compressibility* $(cm^2/dyn).10^{12}$	Θ (mN.m/mg)
1	F-Actin	-6.30	41.3
2	Bovine serum albumin	10.50	37.4
3	α-Casein	5.68	16.3
4	β-Casein	3.80	14.3
5	Ovotransferrin	4.89	33.1
6	Cytochrome C	0.006	13.3
7	Lysozyme (hen)	4.67	23.0
8	Fibrinogen	-	13.2
9	Ovoglobulins	-	34.0
10	Soy 11S globulin	-	30.3
11	Hemoglobin	10.90	61.8
12	α-Lactalbumin	8.27	38.6
13	β-Lactoglobulin	8.45	42.0
14	Lipase	-	29.3
15	Myosin	-18.00	30.9
16	Ovalbumin	8.78	31.3
17	Ovomucoid	3.38	24.0
18	Ribonuclease A	1.12	16.0
19	Tropomyosin	-41.00	36.0
20	Trypsinogen	1.34	10.1
21	Phage T_4 Lysozyme	-	28.0
22	Lysozyme (human)	-	29.9

* From Ref. 14.

$$\beta_s = 0.21\Theta - 0.279 \qquad [4]$$

It should be noted however that the Θ values of fibrous proteins, such as F-actin, myosin, and tropomyosin, which have negative compressibility values, exhibit no relationship with β_s. The excellent linear correlation between compressibility and Θ values of globular proteins, derived from the slopes of the Γ-Π plots under dynamic adsorption conditions, seems to strongly support the proposed hypothesis that Θ is in essence related to flexibility of proteins. This also demonstrates that molecular flexibility is quintessential to surface activity of globular proteins. By experimentally determining the compressibility of physically, enzymatically, or genetically modified proteins, it should be possible to predict their surface activity and their potential applications in food foams and emulsions. In other words, compressibility can be used as a molecular *descriptor* or as a tool to predict a protein's surface activity. Also, by experimentally determining the

Θ value of a protein at the air-water interface, its compressibility can be estimated from equation [1].

Relationship Between Θ and Amino Acid Composition

Since the flexibility (or conversely the compressibility) of proteins is related to volume fluctuations in proteins, it is positively correlated to the partial specific volume of proteins (7). That is, the larger the partial specific volume, the greater is the compressibility of the protein. The partial specific volume of proteins is the sum of three components, i.e.,

$$\upsilon^\circ = V_c + V_{cav} + \Delta V_{sol} \tag{5}$$

where V_c is the sum of the atomic volumes, V_{cav} is the sum of the volumes of void spaces in the interior of the protein, and ΔV_{sol} is the volume change due to hydration. Globular proteins with large υ° generally have large void spaces in the interior. Fibrous proteins, which have a large surface area to volume ratio, are mostly devoid of void spaces, and hence their partial specific volume is low and consequently they exhibit negative compressibility values on account of predominant hydration effects derived from the hydration of the protein surface. Void spaces in globular proteins mainly arise from imperfect packing of hydrophobic residues in the protein interior. Conversely, highly hydrophobic globular proteins must be more compressible than the less hydrophobic globular proteins. This suggests that a fundamental relationship must exist between Θ values and the amino acid composition of proteins. We made attempts to express the Θ value of a protein as a sum of contributions from various amino acid residues. That is,

$$\Theta_j = \Sigma \, n_i \theta_i \tag{6}$$

where n_i is the mole fraction of the ith type amino acid residue and θ_i is the partial specific Θ value of the ith type amino acid. The θ_i values (unknown variable) of 18 amino acids (Glu +Gln and Asp+Asn were grouped as Glx and Asx, respectively) in proteins were determined from multiple regression of 21 equations representing 21 proteins (j=21) for which the Θ values (known variable) were experimentally determined. The partial specific θ value of each amino acid determined in this manner is given in Table II. It should be noted that while the partial specific θ values of a majority of nonpolar amino acid residues are positive, the values of Arg, Lys, Cys, Phe, Pro, Ser, and Tyr are negative. The negative value for Cys, which exists predominantly as cystine in proteins, is reasonable because intramolecular disulfide bonds would hinder unfolding of proteins at interfaces. The positively charged residues Arg and Lys may impart electrostatic repulsive interactions and thereby interfere with formation of a cohesive film. It is interesting to note that proline negatively contributes to Θ of proteins. This is quite unexpected because proline is a helix-breaker and supposed to promote flexibility of polypeptide chain. It is probable that since the dihedral angle ϕ of the N-C$_\alpha$ bond of proline residues in proteins has a fixed value of 70°, it may probably introduce steric hindrance for cohesive interactions between polypeptide segments at the interface. Thus, proteins with a high proline content may not be able to form a cohesive film even if they are flexible. As discussed earlier, this might be one of the reasons for poor cohesive interactions in β-casein (which has a proline content of 17%) films.

Among the amino acid residues, histidine has the highest partial specific θ value of 11.28 mN.m/mg, suggesting that proteins with a high histidine content should be able to form an extremely cohesive film. Since the pK of histidine residues in proteins is about 7.0, at least half of the histidine residues in proteins are uncharged at neutral pH. In the uncharged state, the polarity of the histidine chain is much higher than the other uncharged polar residues in proteins, owing to two nitrogen atoms in the imidazole ring. This may promote polar interactions between segments of proteins at the air-water interface.

Table II. Partial Specific θ Values of Amino Acids

Amino Acid	$\theta\ (mN.m.mg^{-1})$	Amino Acid	$\theta\ (mN.m.mg^{-1})$
Alanine	2.23	Leucine	1.68
Arginine	-2.20	Lysine	-2.96
Aspartic acid+		Methionine	1.37
Asparagine	0.74	Phenylalanine	-1.77
Cysteine	-0.26	Proline	-2.19
Glutamic acid+		Serine	-3.45
Glutamine	1.11	Threonine	1.23
Glycine	0.94	Tryptophane	2.12
Histidine	11.28	Tyrosine	-2.84
Isoleucine	0.33	Valine	2.61

The Θ values of 21 proteins calculated from the partial specific θ values are plotted against the experimentally determined Θ values in Figure 5. The regression line with a slope of close to one (i.e., 0.981) indicates excellent agreement between the experimental and calculated values of Θ. In other words, using the partial specific θ values of amino acids given in Table II and from the knowledge of the amino acid composition, one can predict fairly accurately the Θ value of a protein at the air-water interface. This Θ value can be used as an index of surface activity of the protein and also to predict the compressibility or flexibility of an unknown protein using equation [4].

Relationship Between Θ and Secondary Structure. Intuitively, the flexibility of a protein must be related to the distribution of various secondary structural elements. If Θ is related to flexibility, then it must also be related to the secondary structure contents of the protein, because proteins containing a high amount of β-sheet structures are generally more rigid than those containing relatively high α-helix and aperiodic structures. Theoretically, the Θ can be expressed as

$$\Theta_j = f_h \Theta_h + f_b \Theta_b + f_t \Theta_t + f_{rc} \Theta_{rc}$$ [7]

where f_h, f_b, f_t, and f_{rc} are the fractions of α-helix, β-sheet, β-turns, and random coil contents, respectively, of the protein and Θ_h, Θ_b, Θ_t, and Θ_{rc} are the partial specific θ values of α-helix, β-sheet, β-turns, and random coil structures. The partial specific θ values (unknown variables) of the four secondary structure elements of proteins were determined from multiple regression of 10 equations representing 10 different proteins (j=10) for which the secondary structures contents are available and the experimental Θ values (known variable) have been determined. The least-squares multiple regression of the 10 proteins with a correlation coefficient of (0.98) was

$$\Theta = 0.924 f_h + 0.366 f_b - 1.71 f_t + 0.448 f_{rc}$$ [8]

The Θ_{St} values calculated from this equation are plotted against the experimental Θ values in Figure 6. Considering the fact that other structural factors might also impinge on surface activity of proteins, a slope of close to one (i.e., 0.932)for the regression line shown in Figure 6 indicates excellent agreement between the two Θ values.

Equation [8] suggests that among the secondary structural elements, α-helix has the greatest impact on the surface activity of proteins, followed by the aperiodic structure; the β-sheet contributes only marginally and the β-turn acts as an antagonist. Based on these findings, one should expect that α-type proteins should be more highly surface active than β-type proteins. This is in fact borne out by the fact that BSA, which contains about 60% α-helix and the remainder aperiodic structure, is more surface active than soy 11S globulin, which contains about 64% β-sheet, 9% α-helix, and 27% aperiodic structures (8). The positive impact of α-helix on surface activity is mainly attributable to the fact that it is amphiphilic in nature. That is, one-half of the helix surface is hydrophobic and the other half hydrophilic. Because of this ideal distribution of hydrophilic and hydrophobic surfaces, the α-helix can lie flat on the air-water or oil-water interface, with the hydrophobic surface toward the nonpolar phase and the hydrophilic surface toward the aqueous phase. In addition, as mentioned earlier, the thermal stability of α-type proteins is generally lower than that of β-type proteins. If so, it is reasonable to expect that α-type proteins would unfold to a greater extent than β-type proteins at interfaces.

Since the experimental Θ values were independently correlated with the Θ values calculated from partial specific θ values of amino acids ($\Theta_{Cal, AA}$) and the partial specific θ values of secondary structures ($\Theta_{Cal, St}$), then $\Theta_{Cal, AA}$ and $\Theta_{Cal, St}$ values also must be correlative with each other. This is shown in Figure 7. A slope of close to 1 for the regression line shown in Figure 7 indicates excellent agreement between these values, which validates the contention that the partial specific θ values presented here (Table II and equation 8) can be used to predict surface activity of an unknown protein.

Correlation of Θ with Surface Hydrophobicity. Kato and Nakai (9) showed that the emulsifying properties of proteins were related to the hydrophobicity of the proteins' surfaces as measured by binding of fluorescent probes such as cis-parinaric acid or 1-anilinonaphthalene-8-sulfonic acid (ANS). A plot of experimental Θ values versus

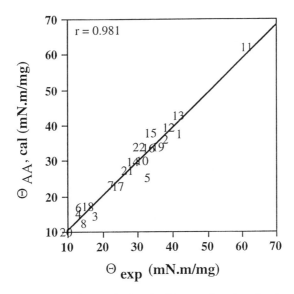

Figure 5. Plots of Θ values calculated with equation 6 and using the partial specific θ values of amino acids (Table II) against the experimental Θ values. The numbers of the data points refer to the identity of proteins in Table I. The solid line represents the least-squares linear regression equation $\Theta_{AA, cal} = 0.965\Theta_{exp} + 0.92$

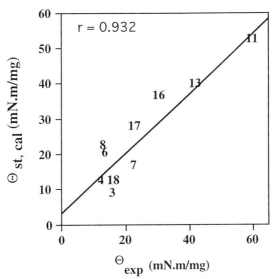

Figure 6. Plots of Θ values calculated with equation 8 against the experimental Θ values. The numbers of the data points refer to the identity of proteins in Table I. The solid line represents the least-squares linear regression equation $\Theta_{St, cal} = 0.852\Theta_{exp} + 3.11$

surface hydrophobicities, as determined by the *cis*-parinaric acid binding, of 21 proteins is shown in Figure 8. Apparently, there seems to be no correlation between surface hydrophobicity and Θ of proteins. This is reasonable because surface hydrophobicity may probably be important only for facilitating initial anchoring of the protein at the interface; it may not be relevant for subsequent cohesive interactions that occur during film development, which involves conformational changes and interaction between the newly exposed functional groups. Thus, it is not surprising that the Θ value, which is related to conformational flexibility and cohesive interactions between protein molecules, is not correlative with surface hydrophobicity.

Correlation of Θ with Emulsifying and Foaming Properties. Figure 9 shows the relationship between Θ and the emulsifying activity index (EAI) of proteins. The EAI was determined by the turbidometric method of Pearce and Kinsella (10). A reasonably linear correlation, with a correlation coefficient of 0.76, is found between EAI and Θ.

The relationship between Θ and the foaming properties of proteins reported in the literature (11, 12) are shown in Figure 10A and 10B. Both foaming power and foaming capacity, determined by using two different methodologies, showed excellent linear relationship (r >0.8) with Θ. A larger data base involving many different food proteins may provide a much better correlation, which might be useful in predicting the foaming properties of unknown proteins or genetically engineered food proteins from experimentally determined Θ values. Since only microgram to milligram quantity of protein is required to perform adsorption studies, this approach can be used as a screening tool in protein engineering.

Correlation Between Θ and Protein Digestibility. It is known that the structural state of a protein affects its digestibility by proteases (13). Native proteins, especially plant proteins, are generally less susceptible than partially denatured ones. It is also reasonable to assume that proteins that are less stable might be more digestible than those that are highly stable. Because, with less stable proteins, initial cleavage of a few peptide bonds by proteases might induce a strong cooperative unfolding of the protein and thereby further hydrolysis by the proteases. Since the Θ value is hypothesized here to be related to flexibility of proteins, it is likely that it might be also correlative with proteolytic digestibility of proteins.

The relationship between Θ and the initial rate of digestibility of various proteins by site-specific trypsin and α-chymotrypsin is shown in Figure 11. As predicted, the rate of digestion shows a strong correlation (r = 0.87) with Θ, further supporting the contention that Θ is a measure of the molecular flexibility of proteins.

Conclusion

The data presented here show that the single most important molecular property that impacts surface activity of proteins is the molecular flexibility. The compressibility of proteins, which reflects molecular flexibility, can be used as a molecular descriptor to assess surface active properties of proteins.

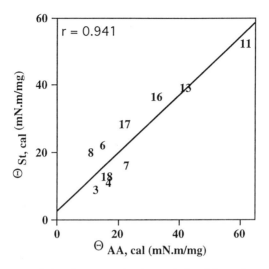

Figure 7. Cross correlations between Θ values calculated from the partial specific θ values of amino acids (equation 6 and Table I) and Θ values calculated from secondary structure content (equation 8). The numbers of the data points refer to the identity of proteins in Table I. The solid line represents the least-squares linear regression equation $\Theta_{St, cal} = 0.86\Theta_{AA, Cal} + 2.63$

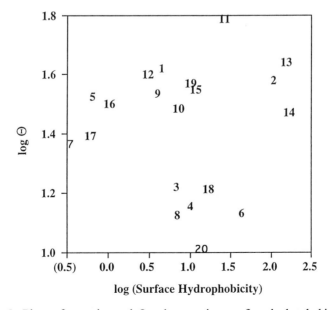

Figure 8. Plots of experimental Θ values against surface hydrophobicity of proteins. The numbers of the data points refer to the identity of proteins in Table-I.

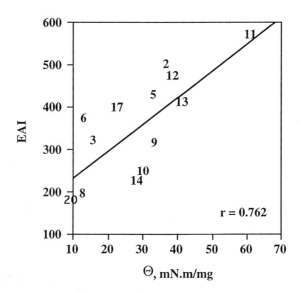

Figure 9. Plots of experimental Θ against emulsifying activity index of proteins. The numbers of the data points refer to the identity of proteins in Table I. The solid line represents the least-squares linear regression.

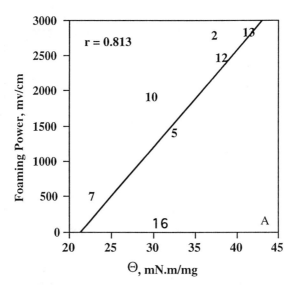

Figure 10. Plots of experimental Θ against foaming power (A) and foaming capacity (B) of proteins. The numbers of the data points refer to the identity of proteins in Table I. The solid line represents the least-squares linear regression. The foaming power and foaming capacity values were taken from references 11 and 12, respectively.

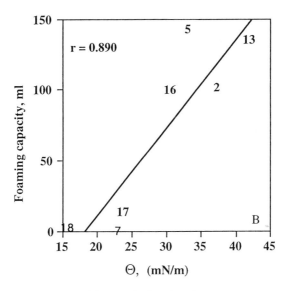

Figure 10. *Continued.*

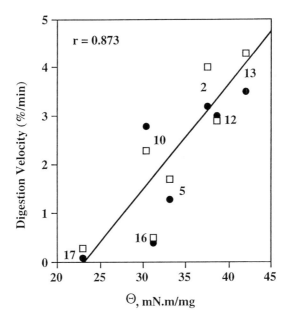

Figure 11. Relationships between experimental Θ and rates of digestion by α-chymotrypsin (□) and trypsin (●). The numbers of the data points refer to the identity of proteins in Table I. The solid line represents the least-squares linear regression.

18

References

1. Miller, S.; Janin, J.; Lesk, A.M.; Chothia, C. *J. Mol. Biol.* **1987**, *196*, 641-656.
2. Damodaran, S. In *Food Proteins and Their Applications*; Damodaran, S.; Paraf, A., Eds.; Marcel Dekker, New York, 1997; pp.57-110.
3. Xu, S.; Damodaran, S. *J. Colloid Interface Sci.* **1993**, *157*, 485-490.
4. Davies, J.T.; Rideal, E.K. *Interfacial Phenomena*, 2nd ed., Academic Press, New York, 1963.
5. Graham, D.E.; Phillips, M.C. *J. Colloid Interface Sci.* **1980**, *76*, 240-250.
6. Cooper, S. *Proc. Natl. Acad. Sci.*, USA, **1976**, *73*, 2740-2741.
7. Gekko, K.; Hasegawa, Y. *Biochemistry* **1986**, *25*, 6563-6571.
8. Wang, C.-H.; Damodaran, S. *J. Agric. Food Chem.* **1991**, *39*, 433-438.
9. Kato, A.; Nakai, S. *Biochim. Biophys. Acta* **1980**, *624*, 13-20.
10. Pearce, K.N.; Kinsella, J.E. *J. Agric. Food Chem.* **1978**, *26*, 716-722.
11. Kato, A.; Komatsu, K.; Fujimoto, K.; Kobayashi, K. *J. Agric. Food Chem.*, **1985**, *33*, 931-934.
12. Townsend, A.; Nakai, S. *J. Food Sci.* **1983**, *48*, 588-594.
13. Deshpande, S.S.; Damodaran, S. *J. Food Sci.* **1989**, *54*, 108-113.
14. Gekko, K. In *Water Relationships in Food*; Levine, H; Slade, L., Eds.; Plenum Press, New York, 1991; pp. 753-771.

Chapter 2

Computer-Aided Optimization of Site-Directed Mutagenesis of *Bacillus stearothermophilus* Neutral Protease for Improving Thermostability

S. Nakai, S. Nakamura[1], and M. Ogawa

Department of Food Science, University of British Columbia, Vancouver, British Columbia V6T 1Z4, Canada

A graphical solution using Random-Centroid Optimization (RCO) was successful in finding the global optimum in multimodal functions. The RCO program was modified to apply to site-directed mutagenesis (RCG). This RCG program was used to mutate one site at-a-time in the active-site helix region (G139-Y154) of *B. stearothermophilus* neutral protease to improve thermostability. Mutant V143E increased the half-survival temperature (T_{50}) by 6.3 °C (T_{50} of the wild-type enzyme = 68.3 °C) after 13 mutations. Meanwhile by the proline mutant (I140P) increased the T_{50} by 7.5 °C. The increase in stability in both mutations was explained by molecular rigidification. Mutant I140P also increased the yield of thermolysin-mediated synthesis of Z-L-Asp-Met-OMe by a factor of 5. The RCG program written for mutating up to 3 sites may be useful in elucidating the structure-function relationships of proteins.

According to Sander (*1*), nature has evolved highly intricate and useful proteins over many millions of years, gradually optimizing protein function in response to selective pressure. The design of proteins or peptides with novel functions can be achieved either by modifying existing molecules or by inventing entirely new structures and sequences that are unknown in nature. Breaker and Joyce (*2*) discussed two major strategies for generating novel biological catalysts as rational design and iterative selection method (irrational design). Generating a pool of random or partially randomized nucleic acids, and then carrying out repeated rounds of selective amplification would yield RNA and DNA molecules, which bind with high affinity to a target protein or small molecules. These two strategies are best applied in a complementary fashion.

[1]Current address: Department of Food and Nutrition, Ube College, Ube, Yamaguchi-ken, Japan 755.

The above situation is similar to the process for developing sequential optimization techniques, the so-called Evolutionary Operation (EVOP) techniques. The EVOP techniques, as reviewed by Lowe in 1964 (3), include Box EVOP, rotating square EVOP, random EVOP and simplex EVOP. Simplex optimization became an important algorithm in the sequential optimization technology (4). With the exception of the random EVOP, some mathematical rules control the EVOP technique. They are, therefore, mostly rational designs.

Schwefel (5) stated that the most reliable global search method is the grid method, which is also the most costly method due to a large number of experiments required before homing in on the global optimum. He elaborated that random search is simple, flexible and resistant to perturbations. However, the method is often inefficient. The introduction of some deterministic rules instead of being probabilistic may however avoid this problem. Nakai et al. (6) have written a Windows program of the Random-Centroid Optimization (RCO) method. The basic unit of RCO is a search cycle consisting of random design, centroid search and mapping. Since theoretically, the random search does not follow specific rules, it is basically an irrational design.

It is interesting to compare the above approach as an optimization strategy with the natural genetic sequence (evolution) which consists of heredity and mutation. Mutation in biology is a sudden departure from the heredity caused by a change in a gene or a chromosome. Because of the irrational nature of mutations, random mutations are customarily used as an important step in genetic engineering to mimic the Darwinian evolution. Evolution of living things may, however, not be a purely random process. Therefore, the biological evolution is slightly different from that seen in mathematics, such as EVOP. EVOP is rather rational, while biological evolution is a mixture of rational (heredity) and irrational (mutation). Precisely, mutation should be classified as irrational due to lack of full explanation of its mechanism.

The above situation is similar to the mathematical random strategy where it is common to resort to random decisions during optimization whenever deterministic rules do not have the desired success, or lead to a dead end (5). Random designs, as an optimization strategy, are inefficient because they rely solely on luck or chance, therefore, some regularization is necessary. The search cycle of RCO, which consists of a regularized random design, a central search around the best response, and mapping, thus could be an appropriate algorithm for optimization of site-directed mutagenesis.

Since proline has the smallest degree of conformational freedom of all of the amino acids, a polypeptide chain at a proline residue has appreciably less conformational freedom than at any other residue (7). This characteristic property of proline can be used to stabilize protein structure. However, the location of proline introduction should be carefully chosen so that the new residues should neither create volume interference nor destroy stabilizing noncovalent interaction (8).

The objectives of this paper were to show the results from computer-aided optimization of site-directed mutagenesis of *Bacillus stearothermophilus* neutral protease in order to improve its thermostability, and to compare these results with those of proline introduction in the same region of amino acid sequence of the same enzyme.

Experimental Procedures and Results

Random-Centroid Optimization of Multimodal Functions. The search unit of RCO is a cycle consisting of random search, centroid search and mapping. The random design used in the RCO is not totally random; it is a regulated random design by eliminating localized designs that are apparently useless for the purpose of broad search for the global optimum (6). The centroid search is a search around the best response within the preceding random search (9). This rather localized search, but not the localized design resulting from randomization, was extremely useful in improving the optimization efficiency without wasting useful information already obtained near the best response (9).

Mapping is an approximation of the response surface thus visualizing the progress of optimization operation. Trend curves are drawn on maps by linking data points, which belong to the same subdivisions when the search spaces of factors are divided into three equal subdivisions (Figure 1). By reading maps, the search spaces can be narrowed for use in the subsequent cycle. In the case of Figure 1, the search space for Factor 2 can be narrowed into about half the level range on the X-axis so as to include data points 1, 8 and 4. In the past, there were an inadequate number of trend lines when four subdivisions were used for each factor. This was especially true if the number of factors was large, or in early cycles with a small number of data points. An example was shown in the paper of Lee et al. (10), where not even a single trend line could be constructed on the maps for the 7-factor optimization. Repeated recycling of the search cycle using the search spaces narrowed based on the trend appeared on maps will lead the search toward the true optimum, as long as the definition of the narrowed search spaces are appropriate. Although maps show only a trend and not exact traces or sections of the true response surface, the trends depicted on maps are adequate for guiding the search toward the global optimum.

It was found that by ignoring one or two factors during computation to draw trend curves, the presence of optimum(s) other than the current best response, if any, could be detected. This "intensified line drawing process" is a powerful tool for global optimization of multimodal functions. A mapping process for automatically eliminating one or two factors in sequence was included in the RCO program. This eliminates the labor-intensive manual elimination of different factors and their combinations. If more than two factors are to be eliminated, however, manual elimination is compulsory.

An example of effects of the intensified line drawing is shown in Figure 2 indicating the maps in Cycle 1 (Figures 2B and C) for the maximization of an unconstrained 6-factor trimodal function. We found that most multimodal 6-factor functions found in the literature were constrained, therefore, making it difficult to randomize the optimal factor level values. Randomization was necessary to make the location of global optimum totally unknown to the users during optimization of model functions. The following unconstrained 6-factor functions was, therefore, constructed according to the method of Bowman and Gerard (11):

22

HYPOTHETICAL GRAPHS FOR EXPLAINING MAPPING PRINCIPLE (MAXIMIZATION)

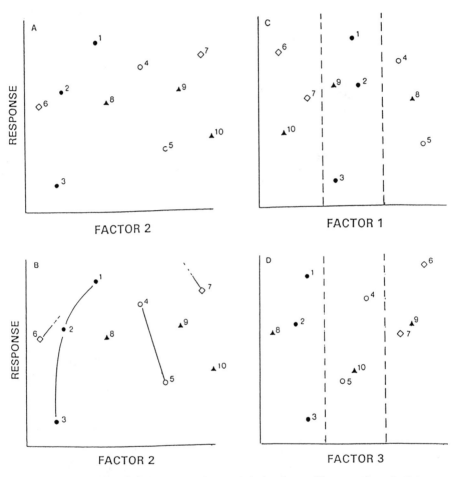

Figure 1.   Maximization in a 3-factor optimization to illustrate the principle of mapping.
A: Before drawing trend curves on scattergram of Factor 2. The location of the maximum is unclear. B: After mapping. The trend lines are pointing toward the maximum. C and D: 3-equal subdivisions of Factors 1 and 3 to find groups of data points that are common in the subdivisions of both Factors 1 and 3. The data points, which fall in the same subdivisions (Figures 1 C and D), can be linked to draw trend lines on the map for Factor 2, as shown in Figure 1B.

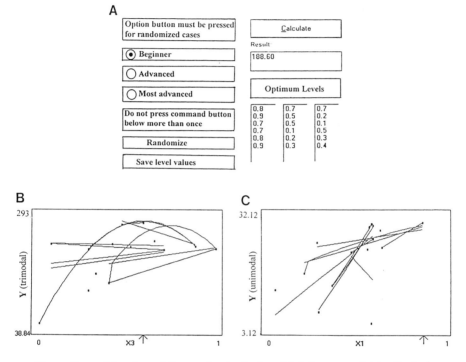

Figure 2. Mapping of the randomized trimodal 6-factor model in Cycle 1 search.

A: The program for response value computation. Three randomized levels of X_1, X_2, X_3, X_4, X_5 and X_6 in Equation 1are appeared under Optimum Levels. Three maxima are combined in a ratio of 5:3;3 with the global optimum of 188.60 as shown under Result. Beginner's randomization was used to avoid unnecessary overlapping with the second and third peaks.[6] B: X_3 is an assembly of peaks at level values of 0.7, 0.5 and 0.1. C: X_1 map of unimodal 6-factor function. The arrow underneath each map points the location of the highest response value.

$$Y = 33.6X_1 + 25X_2 + 41.4X_3 + 10.8X_4 + 8.4X_5 + 10.4X_6 - 4X_1X_2 - 12X_1X_3 - 10X_1X_4$$
$$- 4X_1X_5 - 16X_2X_3 - 12X_2X_4 - 6X_2X_5 - 10X_3X_6 + 16X_4X_6 - 2X_5X_6 - 15X_1^2$$
$$- 18X_2^2 - 20X_3^2 - 26X_4^2 - 10X_5^2 - 6X_6^2 \qquad (1)$$

Figure 2B shows that the presence of three peaks at around 0.1, 0.6 and 0.75. This model is an assembly of three maxima at 0.7, 0.5 and 0.1 of X_3 as shown in Figure 2A under Optimum Levels. However, this does not mean that three peaks appear exactly at the same position of each peak in the mixed function, as the optimum points may slightly shift by combining the peaks. In contrast, Figure 2C is unimodal as shown in the first peak on the extreme left in Figure 2A under Optimum Levels. All trend lines in the X_1 map point toward or above 0.8 even after using the intensified line drawing (Figure 2C). The single peak is supposed to be at $X_1 = 0.8$ as shown in Figure 2A under Optimum Levels.

In addition to this trimodal 6-factor function (Equation 1), the RCO was successful in finding the global optimums for a trimodal 3-factor function of the steep-sided valley of Fletcher and Powell, the bimodal 4-factor Wood's function, the 6-factor Heese's function with 18 local minima, and the 2-factor Curtis' function with several local minima (6). More than 10 randomized optimizations were conducted for each function. While we have found the global optimums within 50 runs in each function for optimization computation, most of the advanced global optimization algorithms, which were computational optimization, consumed sizable CPU time. For example, the number of iterations required for Wood's and Heese's functions were 50824 and 14064, respectively, using the Level Set Program (12).

Optimization of Site-Directed Mutagenesis. The RCO program was modified for site-directed mutagenesis for up to 3-site mutations (13). The modified program (RCG) was then applied to one-site mutation of the active site helix of *Bacillus stearothermophilus* neutral protease in order to improve its thermostability.

Modification of RCO for Mutagenesis (RCG). Two factors are needed for altering a site in the sequences of protein molecules, i.e. the site to be mutated and amino acid to replace the one at the site selected. Since the optimization of multimodal functions up to six factors was successful (Figure 2), the RCG program could be used to accommodate mutations of up to three sites simultaneously. Factors with odd numbers, i.e. factors 1, 3 and 5, were assigned to site locations to be mutated, and factors with even numbers, i.e. factors 2, 4 and 6, were designated for amino acids to substitute for those at the sites selected.

To choose an amino acid for substitution, the hydrophobicity scale of Wilce et al. (14) was used because it was the most recent, thorough investigation of hydrophobicity. In addition, the helix and strand propensities of Muñoz and Serrano (15) and the bulkiness of Gromiha and Ponnuswamy (16) were also included in the RCG program. The propensity for turn or loop was not included, but can be accommodated whenever a reliable scale becomes available. The greatest advantage of including these scales can be found when maps are drawn after each search cycle as functions of different scales, as shown in Figure 3. If trends appear on the maps, this would suggest that those properties of amino acids used in the maps are contributing to the mechanism of the function being optimized. Any scale can be used to select amino acids to substitute for the ones at the sites in question. The RCG program also allows changing the scale in different cycles even during optimization process, if necessary.

Application of RCG to the Active-Site Helix of *B. stearothermophilus* Neutral Protease.

The active-site helix (G139-Y154) was mutated one site at-a-time (*13*). *B. stearothermophilus* neutral protease contains 319 amino acid residues. The results are shown in Table I and Figure 3. The mutant V143E (site 5 in 1-16) gave the greatest increase in the half-survival temperature (T_{50}), i.e. an increase of 6.5 °C as compared to the wild-type enzyme (68.3 °C), with a 30% increase in proteolytic activity (Table I). Figure 3A shows that the mutation of N-terminal end is slightly more effective than at the C-terminal side of the 16-amino acid active site helix. Presence of the active motif HEXXH at sites 7-11 (H145-H149) within helix 1-16 appears as a dip at the middle of the map, which is an example of bimodal function.

Table I. Random-Centroid Optimization of Mutation of Neutral Protease

Optimization step		Site	Mutant	T_{50} (C)*	Proteolytic activity (%)**
Cycle 1	Random	13	V151D	5.7	45.2
		12	A159W	4.4	27.2
		7	H145G	1.7	-87.7
		3	D141P	5.0	76.3
		6	G144F	-0.7	82.9
		10	T148I	2.1	3.6
	Centroid	8	E146.N	3.8	-96.3
		13	V151P	4.2	-12.8
Cycle 2	Random	11	H149W	3.5	-91.5
		7	H145K	6.0	-80.7
		5	V143E	6.5	32.1
		9	L147K	1.9	-1.2
		12	A150E	4.4	21.6

* Difference in T_{50} from that of wild-type enzyme.
** Difference in proteolytic activity from that of wild-type enzyme (100%).

Figure 3B shows the map plotted as a function of bulkiness. The data point, which is on the extreme left, is glycine and is isolated far from the other amino acid data points. Within the bulkiness scale except for glycine, the higher in rigidity the greater in stability. Figure 3C indicates a trend favorable toward low hydrophobicity as demonstrated by the left slanting line as well as data points that are slanting downwards to the right side. This trend may not, however, mean that lower

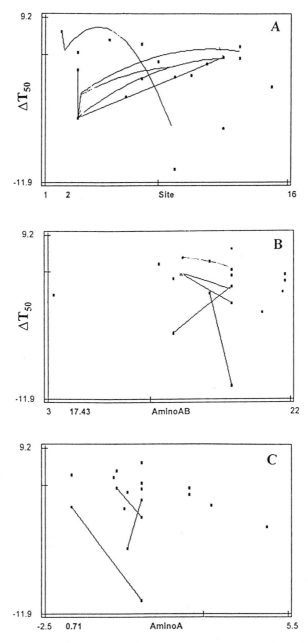

Figure 3.   Maps from random-centroid optimization of mutation of *B. stearothermophilus* neutral protease.

AminoAB: amino acid scale for bulkiness, AminoA: amino acid scale for hydrophobicity, AminoAH: amino acid scale for α-helix propensity, AminoAS: amino acid scale for β-strand propensity, ΔT$_{50}$: half-survival temperature.

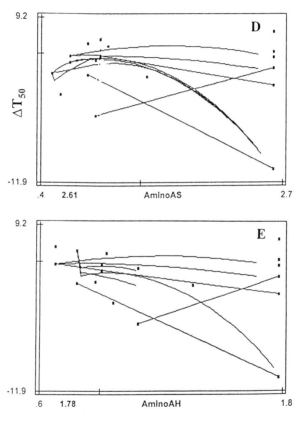

Figure 3. *Continued.*

hydrophobicity is always effective in improving the thermostability of proteins as commonly thought. There is abundant evidence to support an increase in thermostability through replacement with hydrophobic amino acid residues (8). However, the function of proteins is the net result of different factors (17). Furthermore, the large size of aliphatic and aromatic groups in the hydrophobic residues, except charged residues, conflicts with the smaller bulkiness required for thermostability (8). The relative lack of importance of strands and helices is shown in Figure 3D and E, respectively, as there is no definitive trend shown on the maps.

Proline Introduction to the Active-Site Helix of *B. stearothermophilus* Neutral Protease. Two properties of the proline-introduced mutants were investigated, namely thermostability and enzyme-mediated peptide synthesis.

Effects on Thermostability. Despite proline being a helix breaker, introduction of proline residues into α-helices significantly improved the thermostability of human lysozyme (18). This apparent contradiction may be explained by the fact that proline has a unique geometry restricting the conformational freedom of the backbone polypeptide chains (19).

Of 16 amino acid residues (sites 1-16) in the active-site helix region, near the N-terminal sites 2 & 3, middle site 9 and near C-terminal site 15 were replaced with proline (20). Heat stability data of the four mutants and wild type are illustrated in Figure 4. Mutant I140P is the most heat stable with ΔT_{50} of 7.5 °C compared with 2.8 °C, −10.2 °C and 0.5 °C of D141P, L147P and D153P, respectively. α-Chymotrypsin susceptibilities were 77%, 81%, 233% and 105% of wild type for the four mutants in the above order. Rigidification of the molecule can be seen in I140P and D141P, because of the lowest proteolysis value (8).

Table II is the dihedral angles computed from the energy-minimized molecular models. Putative distortion caused by the proline introduction in L147P and D153P is observed as ϕ angle differences of 35° and 11°, respectively, from that of wild type (the bottom half of Table II). A rather dramatic change in L147P may be due to disturbance of the motive HEXXH (sites 145H-149H in Table II) within the active-site helix (21). Absorption of the Pro distortion in the case of I140P and D141P may be due to the presence of Gly ahead of the sites where Pro is introduced as reported by Ueda et al. (22). They concluded that the formation of the Gly-Pro sequence is effective in avoiding possible strain in the folded state of a protein caused by the introduction of proline residue(s). This effect can be seen in Table II. Distortions occurred at the preceding four sites observed in I140P and D141P are less than those in L147P and D153P. Since the importance of the helix coil stability in the thermostability of proteins has been reported (23), the results seen for I140P and D141P mutants as compared to the other two mutants on this helix region may be favorable for maintaining high thermostability to the mutants.

Effects on an Enzyme-Mediated Peptide Synthesis. The mutants, to which proline was introduced, were used for synthesis of an aspartame analog Z-L-Asp-Met-OMe (Nakamura, S.; Nakai, S. *Nahrung*, in press). Yield results obtained by using the four mutants and wild-type in the presence of 50% dimethylformamide are in good agreement with their ΔT_{50} (Table III).

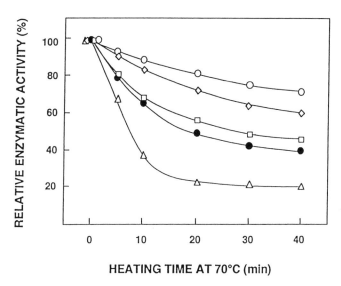

Figure 4. Thermal inactivation of wild-type and mutant neutral proteases. Enzymes in 50 mM Tris-HCl, pH 7.5 containing 5 mM CaCl$_2$ were heated at 70 °C for the residual protease activity using casein as substrate. The residual protease activities are expressed as percent of the 0-time activity. ●, wild-type; ◯, I140P; ◇, D141P; △, L147P; □, D153P.

Table II. Phi and Psi Angles of Amino Acid Residues between Sites 135 and 154 of Wild-Type Enzyme and Mutants I140P, D141P, L147P and D153P obtained from Energy-Minimized Structures

Sequence		135P	136F	137S	138G	139G	140I	141D	142V	143V	144G	145H	146E	147L	148T	149H	150A	151V	152T	153D	154Y
WT	phi	-52.65	-69.77	-90.85	-77.2	173.6	-60.2	-67.6	-69.6	-60.5	-55.2	-58.2	-68.5	-79.6	-70.3	-57.8	-73.8	-61.1	-55.4	-74.2	-72.7
	psi	126.42	-6.45	-4.44	51.8	129.7	-53.1	-27.2	-46.1	-49.1	-42.3	-55.2	-38.4	-11.1	-32.3	-14.5	-56.9	-56.1	-42.8	-34.1	-36.3
I140P	phi	-52.74	-70.06	-90.96	-77.9	177.7	-67.7	-69.2	-71.6	-60.7	-55.5	-58.3	-68.3	-79.8	-70.4	-57.8	-73.9	-60.9	-55.3	-74.1	-72.7
	psi	126.77	-5.76	-1.56	50.8	135.2	-42.8	-30.0	-46.4	-49.1	-41.4	-55.3	-38.5	-11.1	-32.1	-14.3	-57.1	-56.2	-42.7	-34.2	-36.4
D141P	phi	-52.52	-69.99	-90.66	-77.5	173.4	-57.4	-67.8	-69.1	-61.0	-55.6	-58.1	-69.0	-79.6	-70.3	-57.8	-73.7	-60.9	-55.1	-74.2	-72.7
	psi	126.59	-6.17	-4.55	51.5	126.8	-47.4	-30.7	-45.6	-48.9	-44.4	-54.6	-37.9	-11.7	-32.5	-14.5	-57.0	-56.5	-42.8	-34.1	-36.4
L147P	phi	-52.28	-69.12	-91.31	-79.8	173.9	-61.1	-67.1	-67.8	-70.6	-79.4	-54.7	-68.5	-74.3	-69.9	-59.8	-77.5	-59.1	-55.2	-74.7	-72.5
	psi	126.10	-6.38	-2.60	52.2	131.0	-52.3	-28.3	-46.7	-13.7	-46.0	-51.6	-47.7	-4.6	-33.2	-11.7	-57.8	-56.5	-42.5	-34.3	-36.4
D153P	phi	-52.38	-69.83	-90.87	-77.3	173.3	-60.3	-67.7	-70.0	-60.9	-55.6	-58.5	-69.3	-83.0	-69.7	-59.5	-83.8	-60.4	-55.4	-67.2	-70.3
	psi	126.31	-6.41	-4.37	51.9	129.9	-53.1	-26.9	-45.8	-48.8	-41.8	-54.7	-35.8	-10.5	-33.5	-3.5	-54.9	-52.2	-53.1	-29.5	-34.2
Difference from WT																					
I140P	phi	-0.09	-0.29	-0.11	-0.6	4.2	-7.5	-1.6	-2.0	-0.2	-0.3	-0.1	0.2	-0.2	-0.2	0.1	-0.2	0.1	0.1	0.1	0.0
	psi	0.35	0.69	2.88	-1.0	5.5	10.3	-2.8	-0.4	0.0	0.9	-0.2	-0.1	0.1	0.2	0.1	-0.2	-0.1	0.0	-0.1	0.0
D141P	phi	0.13	-0.22	0.19	-0.3	-0.2	2.7	-0.2	0.6	-0.5	-0.4	0.1	-0.5	0.0	0.0	0.1	0.1	0.2	0.3	0.1	0.1
	psi	0.17	0.28	-0.11	-0.3	-2.9	5.7	-3.5	0.5	0.2	-2.2	0.5	0.5	-0.6	-0.2	0.0	-0.1	-0.4	0.0	0.0	-0.1
L147P	phi	0.27	0.65	-0.46	-2.5	0.4	0.9	0.5	1.8	-10.1	-24.2	3.5	0.0	5.3	0.4	-2.0	-3.8	2.0	0.2	-0.5	0.3
	psi	-0.32	0.15	1.84	0.4	1.3	0.8	-0.1	-0.6	35.4	-3.8	3.6	-9.3	6.5	0.9	2.7	-0.8	-0.5	0.2	-0.2	0.0
D153P	phi	0.27	-0.06	-0.02	0.0	-0.3	-0.1	-0.1	-0.4	-0.4	-0.4	-0.3	-0.8	-3.4	0.6	1.7	-10.0	0.7	0.0	6.9	2.5
	psi	-0.11	0.04	0.07	0.1	0.2	0.0	0.3	0.2	0.3	0.5	0.5	2.6	0.6	-1.2	10.9	2.0	3.9	-10.4	4.6	2.2

WT: wild-type; Underline shows the sites where proline was introduced.

The above results may support the hypothesis that peptide synthesis mediated by proteases is the reverse equilibrium of proteolysis. Mutant I140P increased the yield three fold compared to that of wild type, i.e. 15.2% to 47.5%. These values were considerably higher than the best value of 9.1% obtained previously in our laboratory by using thermolysin (*10*). In other words, the dipeptide synthesis using I140P was five times more effective than that by thermolysin.

Table III. Effect of Dimethylformamide (DMF) Concentration on Synthesis of Z-L-Asp-Met-OMe using Wild-Type and Mutant Neutral Proteases

DMF %	Yield %				
	Wild- type	I140P	D141P	L147P	D153P
0	3.1	4.2	3.7	1.5	3.2
10	9.4	15.5	10.3	1.8	12.7
50	15.2	45.7	29.2	1.3	13.1
60	3.2	47.5	31.9	ND[a]	5.1
70	ND	12.0	11.3	ND	2.0

[a] Not detected.

Discussion

RCO Strategy in Experimental Global Optimization. Yassien (*12*) stated that because of the highly nonlinear nature of the equations often involved, multimodal phenomena were quite common with engineering system designs. Many algorithms, e.g. Lipschitz, level-set, simulated annealing, genetic algorithm, have been used for global optimization (*24*). Since biological phenomena are generally more complicated than engineering problems, as shown in Figure 3, with very few available working equations, most of the computational algorithms cannot be applied. Although the RCG is rather empirical, its capacity for a guided search in order to find the global optimum may be an important advantage in usually time-consuming, expensive genetic optimization.

Empirical random search for experimental optimization is not unusual in the optimization study. Low (*3*) discussed the random EVOP in his review on EVOP. The Monte Carlo optimization (*5, 25*) is also a popular random search algorithm. However, they are not very efficient for global optimization because they lack a guiding capacity like the mapping in the RCO strategy. Therefore, the Monte Carlo scheme is rarely used independently for experimental optimization, and rather frequently used in conjunction with other algorithms for global optimization (*26*). In a sense, all algorithms belonging to the EVOP category are attempts to improve the optimization efficiency of random search algorithms

Incorporation of Outside Data into the RCG Optimization. The data from proline introduction used in this study were not the results from the computer-aided optimization. These mutations were conducted based on the hypothesis already proposed in the literature (*18, 22*). It is a well-recognized rule in optimization that any available data relevant to the functions being optimized are preferably incorporated into the optimization data in order to improve its efficiency. Accordingly, Figure 3 was drawn after incorporating the data from proline introduction to the optimization data in Table I. This process greatly aided the interpretation of the maps in Figure 3. The RCO as well as RCG programs are capable of entering additional data already known from other sources into their optimization data. However, for the optimization of a function of proteins for which we do not find homologous sequences in the literature, there may be no other choice than relying on computer-aided optimizations such as the RCG.

Mechanisms of Stability of Thermozymes. Vieille and Zeikus (8) reviewed molecular determinants of protein structural and functional stability. No systematic amino acid substitution is responsible for increased thermostability. However, observations from large groups of enzymes indicate that stabilizing substitutions tend to improve the enzyme's packing efficiency through cavity filling and an increase in core hydrophobicity. These effects may increase the overall enzyme rigidity through α-helix stabilization, electrostatic-interaction optimization and conformational-strain reduction. Our results appear to be consistent with this rule suggested by Vieille and Zeikus (8).

The fact that the structural mechanism for thermostability derived from mapping is the same as that derived from proteolysis susceptibility and molecular modeling could be an advantage of the RCG approach. Conducting additional experiments, especially complicated computer-aided molecular modeling, is laborious and expensive, despite the benefits in enhancing the reliability of the data to support the mechanism of molecular stability. Since the elucidation of structure-function relationships is an important step in optimizing the function of a protein, such as the locations of active and/or binding sites of enzymes, this self-guided explanation of function mechanism using the RCG approach may be an advantage of applying the program for genetic optimization.

We realize, however, that the reliability of propensities of secondary structure of proteins derived from the mapping approach may be restrictive. The characteristic structural effects of the sequence GGP in I140P may not be derived from the mapping process based on the currently available propensity values. The determinant of secondary structure is not only the propensities of amino acids, but also other factors such as tertiary interaction which play a dominant role in molecular folding (27). It is well known that the helix-strand transition readily occurs by heating. Furthermore, the secondary structure of peptide sequences in some proteins may undergo major conformational alterations following tertiary rearrangements induced by pH changes or proteolytic cleavage. These results indicate that the mechanism of protein function predicted by using the propensities of amino acids may not be always true.

Tagaya et al. (28) reported that since the variation of dihedral angle (ϕ) of a prolyl residues is highly limited due to its pyrolidine ring, this residue facilitates the turn structure of a loop and might contribute to the stability of whole protein. They suggested that the replacement of Pro17 with Gly in the glycine-rich region of adenylate kinase would make the turn loose, which is in contrast to that observed in the present study. Hardy et al. (29) reported ΔT_{50} of 5.6 °C for mutant A69P of *B. stearothermophilus* neutral protease. This mutation was conducted in a solvent exposed flexible region, i.e. sites T63-A69. Simultaneous two-site or even three-site mutations using the RCG on broader regions of the molecule may, therefore, become essential to further improve thermostability. Moreover, the propensity of β-turn or loop may become important for elucidating the stability mechanism on maps in the future. An immediate question is what could be expected from the simultaneous two-site mutation of I140P and V143E.

Enzyme-Mediated Peptide Synthesis. Promotion of the dipeptide synthesis by I140P is not a regular function of the neutral proteinase. A ten-fold increase in peptide-ligase activity of thiolsubtilisin (S221C) was already reported using the P225A mutation by Abrahmsén et al. (*30*). They explained this increase by partial relief of the steric crowding resulting from the S221C substitution. In thiolsubtilisin, a cysteine residue replaces the active site serine in a seryl protease subtilisin. Because of differences in the mechanism between seryl proteases and metallo-proteases, the mechanism of enhanced peptide-ligase action seen in the present study is likely different between the two enzyme classes. The active-site helix of metallo-proteases possesses not only catalytic activity but also metal-binding function. At any rate, whether the mutations made were the optimum is unknown for both cases of these enzymes. There has been almost no papers published on enzyme-mediated synthesis of Z-L-Asp-Met-OMe (*10*), and our finding may be useful in food processing for fortification of a nutritionally essential amino acid, methionine. It is interesting to note that *B. stearothermophilus* neutral protease belongs to the thermolysin family of metallo-endopeptidases with 85% sequence analogy with thermolysin (*21*). Effects of these single site mutations, as discussed in this study, are remarkable. These results, such as providing a new nutritional function to a dipeptide sweetener, may suggest the feasibility of creating new functions for novel proteins.

Potential Application of the RCG Program. Currently, many research projects are being conducted in attempts to discover bioactive peptides or proteins from natural resources (The 3rd International Conference of Food Science and Technology on Food for Health in the Pacific Rim, University of California Davis, October, 1997). Only a limited amount of information exits on the mechanism of activities of new proteins or peptides along with a difficulty in finding homologous sequences in the literature. The RCG approach may be useful in these cases for explaining mechanism of functionality, or finding novel functions in newly created mutant proteins, such as hormones, enzymes as well as other bioactive peptides.

Effects of glycosylation on thermostability and inhibitory activity of recombinant mouse cystatin C were investigated in our laboratory by using two expression systems, i.e. *Saccharomyces cerevisiae* and methylotrophic yeast *Pichia pastoris* (Nakamura, S.; Ogawa, M.: Nakai, S. *J. Agric. Food Chem.,* accepted). Polymannosylation of the cystatin in *P. pastoris* was preferable to that in *S. cerevisiae* as it was more heat stable without losing the inhibiting activity of sulfhydryl proteinase papain. Yields of the recombinant cystatin C were 110 mg/L culture of *P. pastoris* compared to 1.3 mg/L of *S. cerevisiae*. The high yield of the new expression system will enable the commercial utilization of this natural, potent antiviral/antimicrobial agent, which is otherwise prohibitively expensive.

The RCO program is posted at the website: www.interchange.ubc.ca/agsci/ foodsci/rco.htm for downloading to any PC computer. Whereas, the RCG program needs further work in order to complete the establishment of the optimization strategy for site-directed mutagenesis.

Conclusion

The random-centroid optimization (RCO), a guided random search, allows for the global optimum to be located in multimodal functions by using a mapping process. When the RCO was applied to site-directed mutagenesis (RCG), the different property indices of amino acids were used to replace the sites selected. Maps as functions of all these indices were drawn to approximate the response surface. This mapping process was useful in elucidating the mechanism of structure-function relationships. Therefore, the RCG approach could eliminate the need for structural information in order to continue the optimization. Mutations of two sites simultaneously in a broader region of large sequences are under investigation

Acknowledgments

The authors are grateful to the Natural Sciences and Engineering Research Council of Canada for Grants to support this study.

Literature Cited

1.   Sander, C. *Trends Biotechnol.* **1994**, *12*, 163-167.
2.   Breaker, R. R.; Joyce, G. F. *Trends Biotechnol.* **1994**, *12*, 268-274.
3.   Lowe, C. W. *Trans. Inst. Chem. Eng.* **1964**, *42*, T334-T344.
4.   Walter, F. H.; Parker, L. R., Morgan, S. L.; Deming, S. N. *Sequential Simplex Optimization;* CRC Press: Boca Ration, FL, 1991.
5.   Schwefel, H. –P. *Numerical Optimization of Computer Models*; John Wiley & Sons: New York, NY, 1981; pp 87-103.
6.   Nakai, S.; Dou, J.; Lo, K. V. ; Scaman, C. H. *J. Agric. Food Chem.* **1998**, *46*, 1642-1654.
7.   Ludscher, R. D. *Food Proteins, Properties and Characterization*; Nakai, S.; Modler, H. W., Eds.; VCH Verlagsgesellschaft: Weinheim, Germany, 1996; pp 23-70.
8.   Vieille, C.; Zeikus, J. G. *Trends Biotechnol.* **1996**, *14*, 183-190.
9.   Aishima, T.; Nakai, S. *J. Food Sci.* **1984**, *51*, 1297-1300 & 1310.
10.   Lee, G. I.; Nakai, S.; Clark-Lewis, I. *Food Res. Int.* **1994**, *27*, 483-488.
11.   Bowman, F.; Gerard, F.A. *Higher Calculus;* Cambridge University Press: London, UK, 1967.
12.   Yassien, H. A. Ph.D. Thesis, University of British Columbia: Vancouver, B.C. Canada, 1993.
13.   Nakai, S.; Nakamura, S.; Scaman, C. H. *J. Agric. Food Chem.* **1998**, *46*, 1655-1661.
14.   Wilce, M. C. J.; Aguilar, M. -I.; Heam, M. T. *Anal. Chem.* **1995**, *67*, 1210-1219.
15.   Muñoz, V.; Serrano, L. *Proteins* **1994**, *20*, 301-311.

16. Gromiha, M. M.; Ponnuswamy, P. K. *J. Theor. Biol.* **1993**, *165*, 87-100.
17. Matthews, B. W. *Ann. Rev. Biochem.* **1993**, *62*, 139-160.
18. Herning, T.; Yutani, K.; Inaka, K.; Kuroki, R.; Matsushima, M.; Kikuchi, M. *Biochemistry* **1992**, *31*, 7077-7085.
19. Branden, C.; Tooze, J. *Introduction to Protein Structure*; Garland Publishing: New York, NY, 1991; pp 13-15.
20. Nakamura, S.; Tanaka, T.; Yada, R. Y.; Nakai, S. *Protein Eng.* **1997**, *10*, 1263 -1269.
21. Jiang, W; Bond, J.S. *FEBS Letters* **1992**, *312*, 110-114.
22. Ueda, T.; Tamura, T.; Maeda, Y.; Hashimoto, Y., Miki, T.; Yamada, H.; Imoto, T. *Protein Eng.* **1993**, *6*, 183-187.
23. Warren, G. L.; Petsko, G. A. *Protein Eng.* **1995**, *8*, 905-913.
24. Horst, R.; Pardalos, P. M.; Thoai, N. V. *Introduction to Global Optimization*; Kluwer Academic: Dordrecht, Netherlands, 1995.
25. Hendrix, D. *Chem. Technol.* **1980**, *10*, 488-497.
26. Marinari, E.; Parisi, G. *Europhys. Lett.* **1992**, *19*, 459-463.
27. Minor, Jr., D. L.: Kim, P. S. *Nature* **1996**, *380*, 730-734.
28. Tagaya, M.; Yagami, T.; Noumi, T.; Futai, M.; Kishi, F.; Nakagawa, A.; Fukui, T. *J. Biol. Chem.* **1989**, *264*, 990-994.
29. Hardy, F.; Vriend, G.; Veltman, O. R., van der Vinne, B.; Venema, G.; Eijsink, V. G. H. *FEBS Letters* **1993**, *317*, 89-92.
30. Abrahmsén, L.; Tom, J.; Burnier, J.; Butcher, K. A.; Kossiakoff, A., Wells, J. A. *Biochemistry* **1991**, *30*, 4151-4159.

Chapter 3

Computer Analysis of Protein Properties

A. Rojo-Domínguez and A. J. Padilla-Zúñiga

Departamento de Química, Universidad Autónoma Metropolitana-Iztapalapa,
Apartado Postal 55-534, 09340 México, D. F.

Proteins play a double role: nutritional and functional. The latter is strongly dependent on its molecular structure and stability. Here, we discuss about the possibilities and limitations of some computer methods to unravel part of the subtle relationship between protein structure and function. After a general overview of current algorithms, including amino acid sequence analysis and the use of molecular mechanics to explore the energy landscape of the protein, we describe a molecular modeling test case and compare the results with crystallographic information.

Proteins have been transformed during evolution producing efficiency in their functions much higher than any device designed by our current technology. These functions include the catalysis at low pressure and temperature conditions (i.e. physiological), with optimum selectivity of reactants, with the capacity of being regulated, and without undesirable secondary products. Because of their functional role, proteins can be used as targets for specific drugs against parasites or defective cells (like cancer tumors), as therapeutic agents for the control of physiological pathologies, and as catalysts in highly specific synthesis or degradation of industrial compounds. On the other hand, knowledge of the structural basis of protein function would permit the rational design of specific ligand molecules that regulate, modify or inhibit the biological activity of the macromolecules. Protein functions owe their characteristics not only to their precise, and rather unstable, three-dimensional structure but also to their specific molecular dynamics and flexibility. For this reason, the complex relationship between structure, function, and stability has been extensively studied from different points of view and with all the available mathematical and computational tools (1). Along their existence, computers have significantly contributed to the study of proteins in a variety of ways. As computing technology develops at exponential rates (about one order of magnitude in speed and data storage capabilities every three years), more precise and robust algorithms can be applied to molecular systems of higher number of atoms. Nevertheless, the gap between the current ability of theoretical calculations and the desired level of precision and understanding needed in biotechnological applications is still broad. A great number of approximations and suppositions shall be made in order to simulate and predict protein characteristics with the different available algorithms; and surprisingly, they work in a limited but useful extent. Two main difficulties are frequently found when using computer methods on protein systems. The first one deals with the interdiscipline. The user needs to immerse in a series of different fields in order to use distinct methods to analyze protein characteristics. Each of those methods will make use of different kind of approximations: statistical, thermodynamic, geometric, topological, biologic, mathematical, kinetic, quantum, and computational, among others. The inherent work involved in that task takes us to the second trouble: It is hard to know what is the actual limit of the method employed (beyond which data is overinterpreted), even when the basis of the method are reasonably understood. A significant amount of the new methods available in World Wide Web servers are somewhat experimental, and one of the purposes of their distribution is to test the quality of the results they yield. Therefore caution in the interpretation is never exaggerated.

It is the purpose of this work to overview some of the methods for extracting information from protein sequences and structures, with a brief description of their fundamentals and limitations. Also we discuss some results from our own predictive experience on different proteins.

Computer Analysis

Needless to say, it is hard to box a series of dissimilar items in a small number of sections. However, the computer methods to analyze proteins could be classified by the kind of input they require (see references *1* and *2* for current methodology). Many of them start from the amino acid sequence of the protein to be examined (*2*), and others need the three-dimensional structure of the molecule, which means to know the Cartesian coordinates of every single atom in the protein. Let us start with the first group of algorithms. What can we obtain from the analysis of a single protein sequence? We can easily calculate the exact molecular weight, the amino acid composition, and the mean residue weight; this later figure is necessary for the interpretation of far UV circular dichroism spectra. It is also possible to detect sequence fingerprints for postranslational modifications (like glycosylation or phosphorilation sites), for cleavage by specific proteases, or for one of the already known active sites (*3*). The importance of such fingerprints is encountered not only in the prediction of functional properties, but also in the detection of important protein features for structural prediction. For example, the detection of a glycosylation site (or an active site) permits to assign some positions of the molecular backbone as solvent exposed. Sequence composition, on the other hand, can yield information on the structural class of the protein (*4*), and on the cellular compartment it belongs to. A special place in this description should be assigned to the methods of secondary structure prediction. They have been developed during the last 25 years and now they can predict the conformation of every residue in a protein with an average confidence level above 70%. In general, the most successful methods do not predict on single sequences but on multiple alignments. The quality of the results frequently depends on the characteristics of such alignments, which are not easy to build. The first step consists in the detection of all known sequences with homology to the protein of interest by searching in databases; then, the alignment process may turn into a harder task than secondary structure prediction itself. In some trivial cases, pairwise alignment can be easily done by hand, but having several hundreds of low identity sequences necessarily requires computer assistance and yields different results depending on the alignment parameter chosen. The user intuition is then the next tool to direct the multiple alignment to results with biological sense. In such alignments, it is desirable to have sequences with both high and low sequence identity. This is because residue changes are more important in similar proteins and residue conservations are more significant among distantly related sequences (5). It is also important to have protein sequences from species that cover a broad part of phylogeny. This adds richness to the information contained in the multiple alignment. Of course, we have no control on the number and identity degree of the amino acid sequences related with our protein of interest, but we are obliged to do our best to find the highest number of related sequences. Statistical analysis can then be done to detect, along the multiple sequence alignment, the conservation or divergence of every position, and also the frequency of fingerprints and insertions (or deletions) of protein segments. Some tertiary contacts sometimes can also be detected which permits to have an idea of the three-dimensional packing of secondary structure elements (6).

Regarding the analysis of atomic coordinates, some tools exist which permit the estimation of solvent accessible areas of molecular surface in an atom by atom basis. Also the molecular volume (7) can be distributed among the constituting atoms to sense the local packing of the structure. From coordinates, geometric features as bond lengths and angles are calculated and compared with those generally observed in proteins, this permits to detect some significant divergences caused by incorrect refinement of X-ray data or special features related to molecular function (8). A very frequent study of protein structure consists in the detection and geometric characterization of its hydrogen bonds. This type of intramolecular interaction stabilizes the secondary structure and it is considered as one of the guides for the assignment of secondary structure to a crystal structure (9). The electrostatic potential, which depends on the atomic electric-charge assignments in every point of the molecule and surroundings, can be calculated from a finite difference method (10). But regarding the atomic interactions in a macromolecule, maybe the most useful tool is a forcefield. Its role is to estimate a relative molecular energy for each conformation. In other words, we are able to know if the rotation of a side chain gives a lower energy state (more stable) or if the distortion from ideality of a bond length or angle is worth enough to optimize a hydrogen bond. If we had a linear A-B-C molecule, the energy function will be of the form: $E = \mathbf{F}(d_{AB}, d_{BC})$, where d_{AB} and d_{BC} are the atomic distances between the indicated atoms. In other words, we could describe any linear molecular conformation as a point in a plane with the independent variables d_{AB} and d_{BC}. For each of those conformations (note that we have an infinite number of them: every positive value is valid for any distance) we could calculate its energy, as far as we know the analytical form of the forcefield function $\mathbf{F}$. Any constant necessary to estimate the conformational energy in **F** is called a forcefield parameter, and the set of those values is known as parameterization, for example CHARMM (*11*). The energy values for each conformation can be plotted on a third axis over the d_{AB} and d_{BC} plane, thus describing a surface or energy landscape (*12*, *13*). Knowledge (or estimation) of the energy dependence on conformation permits to regularize geometry through minimization of energy. This means that we can calculate the direction of the movements of each atom to get a lower energy conformation, and in a series of consecutive steps -maybe several hundred- a molecular geometry in an energy minimum can be obtained. The forcefield is also used to simulate the dynamic behavior of the protein atoms. First, it is assigned random velocities to each atom, and then the force exerted on them by the rest of the molecule is calculated to accelerate the movement. After a simulation period, kinetic energy is distributed on the protein, being the heavier or spatially restricted atoms the slowest and those on the surface or loop side chains the most mobile. It should be kept in mind that these forcefields (frequently referred as empirical forcefields or classical mechanics approximations) consider each atom as a mass and charge point centered in its nucleus. Also, they do not deal with any kind of local polarization or quantum effect, but mainly on a kind of harmonic oscillator model (like a spring) for each type of interaction. In spite of its simplicity, a great number of applications have been found for this approach. For example, molecular modeling or structure predictions rely on the use of energy minimization and frequently on molecular dynamics. Three different types of modeling can be identified: homology (*14*), pattern recognition or threading (*15*, *16*), and *ab initio*. The first two make use of the known crystal structure as a template. The difference is that this structure is detected by sequence homology or by other techniques which measure how compatible is the sequence to be modeled to a database of three-dimensional structures. The *ab initio* method in contrast, tries to obtain structural information only from amino acid sequence and general principles, and it is attempted when no template can be detected. In any case it is necessary to know how to move or sculpt a structure to go from the template-like conformation to that representative of the analyzing sequence. Molecular dynamics can also be used to estimate changes in free energy between two very similar molecules, such as wild type and a single mutant protein (*17*). The change in free energy can then be interpreted in terms of changes in pKa values (*18*) or binding affinities (*19*), among other applications. Also, molecular dynamics can be used to simulate the first stages of protein unfolding (*20*), thus having information on the last steps of folding (*21*), and even to estimate unfolding enthalpies (*22*).

Molecular Modeling of Zeamatin

Zeamatin is a maize 206-amino-acid protein with antifungal activity (*23*, *24*). We modeled its conformation from the crystallographic structure of thaumatin, a sweet-tasting molecule (*25*), with 54% identity to zeamatin. The conservation of eight disulfide bonds between modeled and template proteins (*23*, *25*) permits a higher accuracy in the structural prediction since the covalent S-S bonds anchors different zones of the molecular backbones (Fig 1), thus restricting the number of degrees of freedom. Two domains compose the structure of this family of proteins: the bigger one

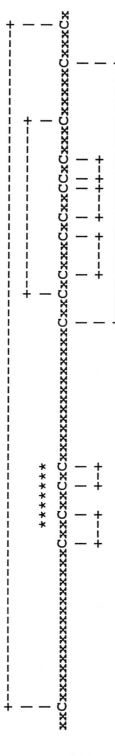

Figure 1. Schematic representation of zeamatin-like proteins. Dotted lines represent disulfide bonds; C, cystein residues; and asterisks the position of the fingerprint of the family.

formed by a β-sandwich, and the smaller by two more strands and a couple of helices. The location of active sites has been proposed to be in the interdomain cleft. Modeling started from the secondary structure prediction of zeamatin and the alignment of its sequence with that of the protein that will guide the three-dimensional construction (Fig 2). More than 80 zeamatin residues, non identical to those in the template, had to be changed *in machina* with the BIOGRAF computer package (Molecular Simulations, Inc., San Diego CA); also, three segments were deleted and five inserted to yield the preliminary model. Each amino acid substitution included the replacement of side chain atoms and the local regularization of conformation in four consecutive steps of energy minimization, molecular dynamics and another energy minimization. First, all atoms were fixed except those of the replaced side chain. Then, in the following stages, regions of a growing number of atoms were allowed to adjust their three-dimensional positions: all atoms of the modified residue, residues at each side of the mutated one, and all atoms in the structure closer than 5 Å to the replaced residue. Since the energy minimization and molecular dynamics in each of the steps were performed *in vacuo*, we used a dielectric constant with a linear dependence on distance in order to simulate the effect of water masking electrostatic interactions. All energy calculations made use of the Dreiding II forcefield (*26*) and a cut off radius of 8.5 Å around each mobile atom. This means that, in order to reduce the demand of computational resources during simulations, only the interactions between atoms located at a distance lower than 8.5 Å were considered. It was also necessary to use a smooth function which slowly "turns off" interactions between 8 and 8.5 Å to avoid discontinuity in the interaction forces inside the interaction sphere. All minimizations were carried out until convergence was reached (forces lower than 0.1 kcal/Å), and the length of each molecular dynamics was 5 ps. Further molecular edition was necessary to delete or insert sequence segments (a total of eight), using soft harmonic forces to open an space for insertions or to close the gap left by a deletion. These forces are added to the forcefield by the user of the package, and can be physically considered as a spring, which exerts a force between two atoms of the molecule. The strength and equilibrium position of the spring is controlled by two parameters given by the user. Geometry regularization is again necessary after all the changes in chain length. Finally, a global molecular dynamics simulation was ran for 100 ps, reaching convergence approximately at 50 ps; the final geometry was the minimized average of the last 50 ps of the simulation, with a 2 ps interval between snapshots.

Results and Discussion

The crystal structure of zeamatin was reported after our modeling, and permitted us to compare the predicted and experimental conformations. The secondary structure of zeamatin was predicted by the PHD server (*27*) and Figure 3 shows the output of the program compared with the β-strands and α-helices found by X-ray crystallography. It can be seen that most of the secondary structure elements were correctly predicted by PHD, albeit some imprecision is found in the detection of their edges. Also, some strands and helices were not detected, most possibly due to the importance of tertiary contacts in the local conformation of those regions. As generally observed, all insertions and deletions are found outside helices and strands, thus avoiding distortion

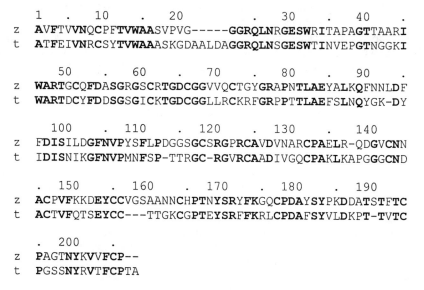

```
         1   .   10    .   20        .   30    .   40     .
z    AVFTVVNQCPFTVWAASVPVG-----GGRQLNRGESWRITAPAGTTAARI
t    ATFEIVNRCSYTVWAAASKGDAALDAGGRQLNSGESWTINVEPGTNGGKI

         50   .   60    .   70    .   80    .   90     .
z    WARTGCQFDASGRGSCRTGDCGGVVQCTGYGRAPNTLAEYALKQFNNLDF
t    WARTDCYFDDSGSGICKTGDCGGLLRCKRFGRPPTTLAEFSLNQYGK-DY

        100    .   110    .   120    .   130    .   140
z    FDISILDGFNVPYSFLPDGGSGCSRGPRCAVDVNARCPAELR-QDGVCNN
t    IDISNIKGFNVPMNFSP-TTRGC-RGVRCAADIVGQCPAKLKAPGGGCND

        .   150    .   160    .   170    .   180    .   190
z    ACPVFKKDEYCCVGSAANNCHPTNYSRYFKGQCPDAYSYPKDDATSTFTC
t    ACTVFQTSEYCC---TTGKCGPTEYSRFFKRLCPDAFSYVLDKPT-TVTC

        .   200    .
z    PAGTNYKVVFCP--
t    PGSSNYRVTFCPTA
```

Figure 2. Amino acid sequence alignment of zeamatin (z) and thaumatin (t). Numbering refers to zeamatin residues. Bold type represents conserved residues in template and model. The insertions and deletions necessary for proper alignment are shown by -.

```
         ....,...10....,...20....,...30....,...40....,...50....,...60
AA       |AVFTVVNQCPFTVWAASVPVGGGRQLNRGESWRITAPAGTTAARIWARTGCQFDASGRGS|
R-X      | EEEEE      EEEEE    EEEEE     EEEEE     EEEEEEEEEE      EE |
PHD sec  | EEEEE      EEE       EE      EEEEE         EEEE           |
Rel sec  |9279965883374125468999623379974787659999862567266623389999965|
prH sec  |0000000000000010000000000000000000000000000000000000000000000|
prE sec  |0589872113676432221000156310016888720000023778511133300000012|
prL sec  |9410026885312457678899733589983111269999875211477755688998877|
SUB sec  |L.EEEELLL..E...L.LLLLLL...LLLL.EEEELLLLLL.EEE.LLL...LLLLLLLL|

         ....,...70....,...80....,...90....,..100....,..110....,..120
AA       |CRTGDCGGVVQCTGYGRAPNTLAEYALKQFNNLDFFDISILDGFNVPYSFLPDGGSGCSR|
R-X      |EEEE             EEEEE    EEEEEEEEE        EEEE            |
PHD sec  |      EEEE      HHHHHHHHH     EEEEEE    E                  |
Rel sec  |2258767413552899991489894551489995488999837432101688998998855|
prH sec  |0000110011100000005689886664200000000000000000000000000000000|
prE sec  |4420111245763000000000000311000000268899983123544421100000022|
prL sec  |5568777643125889994300000124688987211000168663545788999999877|
SUB sec  |..LLLLL...EE.LLLLL..HHHH.HH..LLLLL.EEEEE.L......LLLLLLLLLLLL|

         ....,..130....,..140....,..150....,..160....,..170....,..180
AA       |GPRCAVDVNARCPAELRQDGVCNNACPVFKKDEYCCVGSAANNCHPTNYSRYFKGQCPDA|
R-X      | EE        HHHEE   EE   HHHHH    HHHH         HHHHHHHHH      |
PHD sec  | EE                           EEE     EEEE                 |
Rel sec  |6322034565696222367898898324199548862899877998511110214799 95|
prH sec  |0000112212102431100000001000000000000000001100024444454310002|
prE sec  |1355421110000023311000000356500268874100000000000000000000000|
prL sec  |7534455676797544578898898533488731125899888998745444446899 96|
SUB sec  |L......LLLLLL....LLLLLLLL....LLL.EEE.LLLLLLLLLL........LLLLL|

         ....,..190....,..200....,.
AA       |YSYPKDDATSTFTCPAGTNYKVVFCP|
R-X      |     HHHH EEEE    EEEEE    |
PHD sec  |          EE      EEEEE   |
Rel sec  |7989999973144379976388883 9|
prH sec  |1000000000000000000000000 0|
prE sec  |0000000013466310012688883 0|
prL sec  |8889999976433589977311016 9|
SUB sec  |LLLLLLLLL.....LLLLL.EEEE.L|
```

Figure 3. Secondary structure prediction by PHD server. Numbering and
sequence belongs to zeamatin. PHD sec and SUB sec denote the secondary
structure predicted for each amino acid, and the subset of highest reliability.
Lines marked by prH, prE, and prL represent the prediction strength for
helix, extended and loop structures, respectively, while Rel-sec line is the
reliability of the prediction for each position (9 highest).

of secondary structure. This fact can be used as a rule of thumb to analyze the multiple sequence alignment of a set of proteins. Positions with frequent long insertions or deletions parse the sequence into segments with complete secondary structures. Figure 4 schematically shows the main chain of the model compared with that of the crystallographic structure. On the left side, it can be seen the main structural domain formed by two β-sheets and two prominent loops in the upper and lower regions, the active site cleft is situated in the top of the figure. The backbone comparison yields a remarkable similarity between theoretical and experimental structures, with an RMS distance between equivalent alpha-carbons of 1.6 Å (average distance 1.3 Å with standard deviation of 0.95 Å). This value is comparable with the value of 0.6 Å obtained from the comparison of the two zeamatin-molecule backbones in the unit cell of its crystallographic study (obtained at 2.5-Å resolution). Considering a single figure, it could be said that the alpha carbon trace of the model correctly describes the structure of zeamatin, within X-ray experimental error. But this statement is false because it is based exclusively on an average value, and it does not reflect how larger differences between the predicted and crystal structures arise in several local regions (Fig 5). For example, seven alpha carbons are predicted with coordinates more than 3 Å apart from the experimental positions. This analysis illustrates a common pitfall in the validation of a predictive method, averaging masks details and a global figure might not represent appropriately the whole data set. This does not discard the modeling of the structure, see Figure 4, but calls our attention to the existence of portions of the backbone with significant divergence between the structures. Most of these zones are in loop regions between helices or beta strands, or close to positions where insertions or deletions were performed. The lower accuracy in the determination of conformation of the solvent exposed regions is frequently found in structure prediction since such segments have less interactions and, in consequence, are subjected to less geometrical restrictions (28). Another interesting comparison is the average difference between the template and the structures of modeled and crystal zeamatin, which yields values of 1.16 ± 0.92 and 0.74 ± 0.75 respectively. This means that the model diverged more from the template than from the experimental zeamatin structure, and it is the expected result. Since the zeamatin atoms in the preliminary model have to move from the thaumatin-like conformation to their final positions through a complex energy landscape with several thousands of freedom degrees, it is reasonable to find some of them slightly "lost", trapped in some local energy minimum. If we had found the opposite, it would be interpreted as a model too close to the template possibly as a result of a bad forcefield not able to drive atoms to their new positions. Our new discussion takes us to the question of how well does a forcefield simulate reality. To get a definite answer it should be necessary to experimentally measure the energy of a series of protein conformations and compare them with the theoretical ones. Furthermore, it should be desirable to estimate with an experimental technique all the terms in the energy function for the comparison, since some internal compensation and averaging can occur in the total conformational energy calculation. It is impossible to perform such kind of measurements, but another comparison between theory and experiment can be done instead. During the last 50 ps of the final molecular dynamics simulation, the mobility of each atom can be determined as the RMS of the distance between its instantaneous and its average positions. The higher

Figure 4. Three-dimensional comparison of alpha carbon traces for zeamatin model (black) and crystal structure (gray).

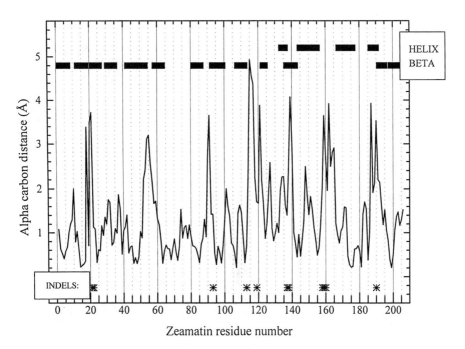

Figure 5. Distance of equivalent alpha carbons between theoretical and experimental zeamatin structures. Upper black bars mark the length and position of secondary structure elements. Positions of insertion or deletion of residues during modeling are shown with asterisks.

the RMS value the higher the mobility of that atom. This value can then be compared with an experimental parameter that measures the spread of the electron density for each atom, the B factor. This value raises not only due to atom mobility in a single molecule but also due to differences between molecules in the crystal. Nevertheless we can consider that two figures for each atom are available, a theoretical and an experimental one, which can be qualitatively compared. Figure 6 shows the value of these variables for the alpha carbon atoms of zeamatin. It is encouraging for predictive methods to find a good correlation between the curves in figure 6, in spite of the fact that solvent molecules were not explicitly included and that almost half of the template side chains should be changed. Up to now, we have found an acceptable agreement between backbone conformation and mobility in our zeamatin model, but most of the functional properties of a molecule are determined by the side-chain solvent accessibility. In contrast to backbone atoms, side chain atoms possess a considerable number of possible conformations. It is then expected a lower precision in their structure prediction, especially for those side-chains situated on the molecular surface. We estimated the maximum solvent-accessible area (*29*) of any side chain as that presented when found in the middle of an extended tripeptide ala-xxx-ala. We then calculated the differences in solvent-accessible area of every side chain between the model and the crystal structure of zeamatin, expressed as a percentage relative to that in the extended tripeptide (Figure 7). Some residues in the model present an accessible surface area higher than in the X-ray crystal structure, and viceversa, but most of them are within a 15% band, with an average value of 9.6 ± 11.2% in their absolute values. With all the caution recommended for an average value, we can consider the model as a good descriptor of the molecular surface of zeamatin, except in some very local zones close to insertion or deletion points. Some other geometrical properties can be compared between the zeamatin structures, for example their stereochemical quality (*30*). The predicted and experimental molecules respectively have 66 and 80 % of their residues in the low energy regions of the Ramachandran map, 6.3 and 9.3 degrees of average deviation from planarity in their peptide bonds, and 12.5 and 18.3 degrees of average deviation from optimal bond angles rotating through the alpha and beta carbon bond (angle chi-1). As a final conclusion of this section we can say that a good model for a protein structure can be constructed from its amino acid sequence and a crystal structure of a template, with more than 30 % of sequence identity. But some local divergence between the predicted and experimentally determined structure can be found in some regions of the molecule, in particular close to the insertion or deletion points which have proven to be hard to model.

Other Predictive Work. Some other cases of homology prediction work have been published by our group, but the above described is the best example since the results can be compared to the recently reported crystal structure. For the case of a lipocalin, the α-1 acid glycoprotein or orosomucoid, it was possible to build a three-dimensional model and to propose a recognition site for progesterone, a known ligand for this protein (*31*). Modeling was difficult due to the low amino acid similarity between the glycoprotein and the closest lipocalin with known structure. The crystallographic conformation of the glycoprotein has not been resolved, possibly due to the microheterogeneity of its carbohydrate moiety. This is a good example of computer

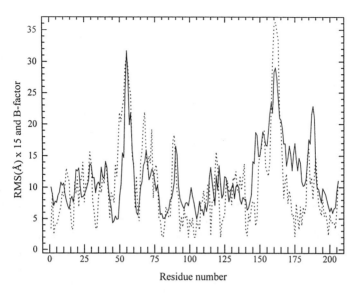

Figure 6. Comparison of crystal mobility (B-factor, dotted line) and the position fluctuation during molecular dynamics simulation (RMS, continuous line). Numbering corresponds to zeamatin.

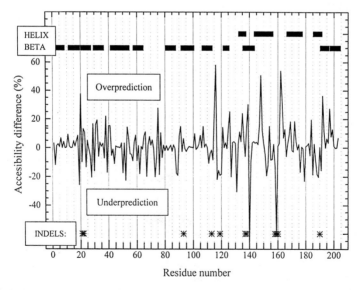

Figure 7. Solvent accessible areas of side chains compared between modeled and crystallographic structure. Positive values correspond to amino acids with higher exposition to solvent in the model than in the crystal structure. Bars and asterisks represent secondary structure and indel positions as in figure 5.

prediction as an alternative to experimental techniques for structure determination. In other studies, computer analysis has helped us in the investigation of the molecular basis of species-specific inhibition of triosephosphate isomerase (TIM), by the comparison of structures from different organisms (*32, 33*). One of the most recent results include the homology modeling of this enzyme from *Entamoeba histolytica*, using the crystal structure of triosephosphate isomerase from yeast. An interesting feature in the amebic molecule is the unusual presence of five amino acid segments not present simultaneously in the rest of the more than 40 TIM sequences currently known. These segments had to be inserted during modeling, and they were found in the same region of the molecular surface but spread through the backbone (*34*). It is possible that such insertions play a role in structure stability and folding. Also, the techniques of molecular modeling can be extended to carbohydrate molecules. We modified the sugar molecule bound to a rubber-tree lectin, hevein, in order to determine structural features and compared them with those of other carbohydrate-lectin complexes. This allowed us to propose that some geometric characteristics in the binding of these types of molecules diverge from the protein-protein interactions, and represent a strong basis for the further understanding of intermolecular recognition (*35*). Currently, we are working on a more ambitious prediction task, the proposal of the folding pattern of a 107-residue protein with no use of crystallographic templates (Padilla-Zúñiga, J.; Rojo-Domínguez, A. *Fold. Des.* **1998**, in press.). In this kind of non-homology modeling the precision is considerably lower than in the other methods, and it makes use of all available experimental data where conformational information can be extracted from. This type of prediction is the only alternative for proteins with sequences without a reliably detectable crystallographic template.

Conclusions

Computer predictive methods for the structural analysis of proteins increase at the same speed as computer technology, numerical algorithms and experimental knowledge do. This is a consequence of the high complexity and interest in the problem. Several hundreds of theoretical groups around the world use all the methodologies in the most sophisticated computer systems trying to unravel one of the most important secrets of Nature for our contemporary science. Although impressive results have been obtained by computer prediction and simulation of protein characteristics, the progress in the field is obtained with extreme effort. The first X-ray protein structure was determined in 1960, and now the Protein Data Bank (*36*) contains several thousands of structures. Most of our knowledge on protein structure comes from these coordinate files and their analysis. Nevertheless computer methods are far to be automatic tools with straightforward outputs. These algorithms should not be considered as black boxes, instead the user must understand the basis of each method and be aware of the limitations of the tool employed. Some data used by software can be only rough estimations of true values and the selection of variable or optional parameters frequently affect the results. For the future, it is reasonable to expect that computers will remain as a powerful tool in this research field, and that in the next millenium we could see a synergistic union of quantum mechanics with computer science which will give us new astonishing results in protein function comprehension.

50

References

1. Böhm, G. *Biophys. Chem.* **1996**, *56*, 1-32.
2. Eisenhaber, F.; Persson, B.; Argos, A. *CRC Crit. Rev. Biochem. Mol. Biol.* **1995**, *30*, 1-94.
3. Bairoch, A.; Bucher, P.; Hofmann, K. *Nucleic Acids Res.* **1997**, *25*, 217-221.
4. Chou, K. C.; Zhang, C. T. *CRC Crit. Rev. Biochem. Mol. Biol.* **1995**, *30*, 275-349.
5. Benner, S. A.; Badcoe, Y.; Cohen, M. A.; Gerloff, D. L. *J. Mol. Biol.* **1994**, *235*, 926-958.
6. Thomas, D. J.; Casari, G.; Sander, C. *Protein Eng.* **1996**, *9*, 941-948.
7. Pontius, J.; Richelle, J.; Wodak, S. *J. Mol. Biol.* **1996**, *264*, 121-136.
8. Morris, A. L.; McArthur, M. W.; Hutchinson, E. G.; Thornton, J. M. *Proteins* **1992**, *12*, 345-364.
9. Kabsch, W.; Sander, C. *Biopolymers* **1983**, *22*, 2577-2637.
10. Gilson, M.; Honig, B. *Nature* **1987**, *330*, 84-86.
11. Brooks, B. R.; Bruccoleri, R. E.; Olafson, B. D.; States, D. J.; Swaminathan, S.; Karplus, M. *J. Comp. Chem.* **1983**, *4*, 187-217.
12. Frauenfelder, H.; Sligar, S. G.; Wolynes, P. G. *Science* **1991**, *254*, 1598-1603.
13. Dill, K. A.; Chan, H. S. *Nature Struct. Biol.* **1997**, *4*, 10-19.
14. Rost, B.; Sander, C. *Annu. Rev. Biohpys. Biomol. Struct.* **1996**, *25*, 113-136.
15. Bryant, S. H. *Proteins.* **1996**, *26*, 172-185.
16. Jones, D. T.; Thornton, J. M; *Cur. Op. Struct. Biol.* **1996**, *6*, 210-216.
17. Karplus, M.; Petsko, G. A. *Nature* **1990**, *347*, 631-639.
18. Alexander, K. D.; Antosiwicz, J.; Bryan, P.; Gilson, M.; Orban, J. *Biochemistry* **1997**, *36*, 3580-3589.
19. Gilson, M. K.; Given, J. A.; Bush, B. L.; McCammon, J. A. *Biophys. J.* **1997**, *72*, 1047-1069.
20. Li, A.; Daggett, V. *J. Mol. Biol.* **1996**, *257*, 412-429.
21. Hinds, D. A.; Levitt, M. *J. Mol. Biol.* **1994**, *243*, 668-682.
22. Lazaridis, T.; Archontis, G.; Karplus, M. *Adv. Prot. Chem.* **1995**, *47*, 231-306.
23. Malehorn, D. E.; Borgmeyer, J. R.; Smith, C. E.; Shah, D. M. *Plant Physiol.* **1994**, *106*, 1471-1481.
24. Batalia, M. A.;. Monzingo, A. F.; Ernst, E. S.; Roberts, W.; Robertus, J. D.; *Nature Struct. Biol.* **1996**, *3*, 19-23.
25. Edens, L.; Heslinga, L.; Klok, R.; Ledeboer, A. M.; Maat, J.; Toonen, M. Y.; Visser, C.; Verrips, C. T. *Gene* **1982**, *18*, 1-12.
26. Mayo, S. L.; Olafson, B. D.; Goddard III, W. A. *J. Phys. Chem.* **1990**, *94*, 8897-8909.
27. Rost, B.; Sander, C. *Proteins* **1994**, *19*, 55-72.
28. Pascarella, S.; Argos, P. *J. Mol. Biol.* **1992**, *224*, 461-471.
29. Hubbard, S. *NACCESS* **1994**, University College London and EMBL Heidelberg.
30. Laskowski, R. A.; McArthur, M. W.; Moss, D. S.; Thornton, J. M. *J. Appl. Cryst.* **1993**, *26*, 283-291.
31. Rojo-Domínguez, A.; Hernández-Arana, A. *Protein Seq. Data. Anal.* **1993**, *5*, 349-355.
32. Gómez-Puyou, A.; Saavedra-Lira, E.; Becker, Y.; Zubillaga, R.A.; Rojo-Domínguez, A.; Pérez-Montfort, R. *Chemistry and Biology* **1995**, *2*, 847-855.
33. Garza-Ramos, G.; Pérez-Montfort, R.; Rojo-Domínguez, A.; Gómez-Puyou, M.T.; Gómez-Puyou, A. *Eur. J. Biochem.* **1996**, *241*, 114-120.
34. Landa, A.; Rojo-Domínguez, A.; Jiménez, L.; Fernández-Velasco, D. A. *Eur. J. Biochem.* **1997**, *247*, 348-355.

35. García-Hernández, E.; Zubillaga, R. A.; Rojo-Domínguez, A.; Rodríguez-Romero, A.; Hernández-Arana, A. *Proteins* **1997**, *29*, 467-477.
36. Bernstein, F. C.; Koetzle, T. F.; Williams, G. J. B.; Meyer, E. F.; Brice, M. D.; Rodgers, J. R.; Kennard, O.; Shimanouchi, T.; Tasumi, M. *J. Mol. Biol.* **1977**, *112*, 535-542.

Chapter 4

Evaluation of Viscous Food Properties Using a Helical Ribbon Impeller

E. Brito-De La Fuente[1], L. M. Lopez[1], L. Medina[1], G. Ascanio[2], and P. A. Tanguy[3]

[1]Food Science and Biotechnology Department, Chemistry Faculty "E", National Autonomous University of Mexico, U.N.A.M., 04510 Mexico, D. F.
[2]Center for Instruments, U.N.A.M., 04510 Mexico, D. F.
[3]URPEI-Department of Chemical Engineering, Ecole Polytechnique Montreal, P.O. Box 6079, Stn. Centre-ville, Montreal H3C 3A7, Canada

The measurement of fluid response under deformation conditions closer to the actual flow process is called mixer viscometry. A helical ribbon impeller fitted to a rheometer was used to measure torque-impeller rotational speed of Newtonian, shear thinning and shear thickening fluids, which simulate many real fluid foods. Raw experimental variables were transformed to process viscosity using laminar-mixing principles. At a given impeller rotational speed, increasing the level of pseudoplasticity, power consumption decreases and thus the process viscosity. Increasing the solid-particle concentration, shear thickening increases. For two-phase fluids, the estimation of the viscous properties with classical geometries produces wrong results. For these systems, a helical ribbon based viscometry gives useful data for design, scale-up and comparison purposes.

Almost all fluid foods of technical and practical importance show rheological complex behavior. In these cases important physical properties such as steady shear viscosity, η, in general, will not be a material function, but rather it will depend upon the stress the fluid is subjected to. The non-homogeneous (i.e. presence of several phases) nature of many food systems add considerable complexity to the experimental evaluation of viscous properties. Blocking of the measuring gap, sedimentation, slip at the wall and phase separation, are common problems which frequently produce wrong rheological results. Therefore, the classical viscometric flows (e.g. Couette flow, cone-plate) have proved in most cases to be unsuitable for these types of fluids (*1,2*).
    An alternative to solve this problem is the measurement of the fluid response under shear stress and shear deformations closer to the actual flow process. The viscous property estimated in this way is also known as process viscosity. In the case of mixing tanks, this is called mixer viscometry (*3*).

The general principle of mixer viscometry is based on measurement of the power draw transferred from the impeller to the fluid. This energy consumption is one of the principal quantities used to describe mixing performance (4). It must be noted here that power input is normally evaluated by measuring the torque on the shaft of the impeller, T, as a function of its rotational speed, N. To extract the viscosity function from the applied rotational speed and the measured torque, the following relationships from mixing theory are used:

1. For Newtonian mixing in the laminar region,

$$K_p = \frac{2\pi T}{Nd^3 \mu} \tag{1}$$

where d is the impeller diameter and μ the Newtonian viscosity.

2. For non-Newtonian fluids following a power law behavior (e.g. The Ostwald-de Waele model: $\tau = m \gamma^n$) and assuming as a characteristic velocity the product Nd,

$$K_p(n) = \frac{2\pi T}{md^3 N^n} \tag{2}$$

where τ is the shear stress, γ is the rate of deformation, and n and m are the flow behavior and flow consistency index, respectively. Equation 2 may also be expressed as:

$$T = \frac{K_p(n) m d^3 N^n}{2\pi} = A(n) N^n \tag{3}$$

where A(n) is a type of shear stress function which for a given mixing system (i.e. impeller-tank geometry) depends on the flow behavior index. From Equation (3), n and m can be determined provided $K_p(n)$ is known. This is possible by performing a previous calibration of the mixing geometry using Newtonian and non-Newtonian fluids of known viscous properties (5). Once n and m are known through Equation 3, a process viscosity curve may be generated.

It is worth noting here that by following the approach described above one may express the fluid viscous response in terms of process variables (i.e. torque and impeller rotational speed). This may be sufficient for comparison purposes (e.g. quality control) but inadequate for design and scale-up of flow processes. In order to use mixer viscometry as a means for rheological testing, some relationships had to be found between the impeller velocity and the shear rates applied to the system, and between the resulting stress and the measured torque. This task has been accomplished in a number of ways in the literature.

The Metzner and Otto correlation (6) is the preferred method to estimate average or effective shear rates, in particular for inelastic shear thinning fluids. Furthermore, this correlation is routinely recommended as a standard procedure in mixing textbooks (7). Although empirically developed, the Metzner-Otto correlation originates from the Couette flow analogy. The basic assumption is that the fluid motion near the impeller can be estimated by an average shear rate, $\dot{\gamma}_e$, which is linearly related to the impeller rotational speed by:

$$\dot{\gamma} = K_s \, N \qquad (4)$$

where K_s is a constant of proportionality dependent, in principle, only on the mixing geometry. Many reports in the literature have suggested that K_s depends on the rheological properties, in particular on the power law index, n (3-4,5,7). For example, for close clearance impellers such as helical ribbon impellers and for highly shear thinning fluids (n < 0.5), K_s depends on n (8).

For mixing applications at industrial scale, in particular for those related with power input estimations, considering K_s as a constant may not impact the design results because of the common use of the so-called "engineering design safety factors". However, for viscometric purposes or for the design of shear sensitive systems, the error on the viscosity estimations may be more important. It is important to mention that most mixer viscometry applications are based on the assumption that Equation 4 is valid and K_s is a constant independent of the fluid rheological properties (3).

In continuation with our past and present work on mixing, it is the intent of this article to study the capabilities of a helical ribbon impeller for viscometric measurements of high precision. We extend here our previous results for homogeneous (i.e. single-phase) shear thinning fluids to heterogeneous (i.e. two phases) shear thinning and shear thickening fluid foods.

Materials and Methods

The mixing-vessel system used in this study is similar to larger scale mixers previously described (8). The cup (vessel)-bob (impeller) was built by keeping constant the geometrical ratios described in Table I.

Table I. Mixing System Geometrical Ratios

Impeller	D/d	h/d	w/d	s/d
HR1	1.17	1.269	0.165	1.269

D = vessel diameter; d = impeller diameter (0.0248 m); h = impeller height; w = ribbon blade width; s = ribbon pitch

Torque and rotational speed measurements for different model fluids and real fluid foods were performed with the vessel-impeller system fitted to a rotational rheometer (Haake CV 20N, Germany). Furthermore, for comparison purposes, steady state shear viscosity functions were determined for all fluids using a standard cone-plate (PK 20/4, Haake) geometry as well as a concentric cylinder or

Couette (ZB-15, Haake) geometry. The whole apparatus was temperature controlled by using a constant temperature bath. All measurements were performed at 25 °C.

Newtonian fluids (e.g., silicon oil, glycerol and corn syrup) with viscosities ranging from 0.1 to 25 Pa s were used for calibration purposes. The model non-Newtonian fluids were formulated using different bio-polymers in several concentrations according to Table II. Single-phase shear thinning fluids were aqueous solutions of common additives used in the food industry. For the two-phase fluids, kaolin in water at different concentrations was used. These dispersions were prepared by adding in all cases 6% wt CMC and 0.9% wt Dixpex (Allied Colloids Inc.,), this last as a dispersing agent. A commercial Mexican salad dressing (Herdez Green Sauce) was also used as a real two-phase fluid food system.

Table II. Fluid Formulations

Fluid	Concentration range (%wt)
Single-phase	
CMC (Sigma Chemicals)	1.5 – 3
Gellan gum (Kelco, Merck)	1 – 1.5
Xanthan gum (Kelco, Merck)	1 – 3
Two-phases	
Kaolin (J.T. Baker)	40 – 60

Results and Discussion

Calibration of the mixing system, in the laminar mixing regime, is expressed numerically by the following single-phase Newtonian and shear thinning power input results (see Equations 1 and 2):

$$K_p = 162.55 \pm 1.04 \qquad (5)$$

$$K_p(n) = 162.55(24.64)^{n-1}(0.91)^{\frac{n-1}{n}} \qquad (6)$$

Equation (6) is a non-linear model previously proposed in the literature (5) which represents quite well power input results, particularly in the high shear thinning region (n < 0.5). The constants of Equation 6 are estimated through a non-linear regression algorithm (5, 8).

In Figure 1, typical results of torque as a function of the impeller rotational speed are shown for a single-phase fluid, 1.5% CMC. The results obtained with the classical cone-plate geometry are also presented for comparison purposes. The representation of the experimental results by Equation 3 is also drawn. From Equation 3, the power law index, n, is calculated for both helical ribbon and cone-plate. This give an n = 0.81 and n = 0.83 for the helical ribbon and cone-plate, respectively. With the value of n and Equations 3 and 6, the value of the

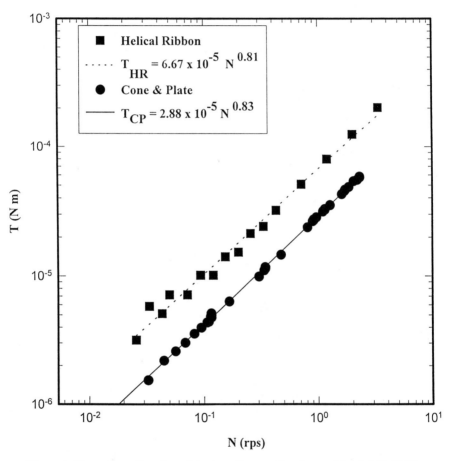

Figure 1. Torque as a function of the impeller rotational speed for 1.5 % CMC (single-phase fluid).

consistency index, m is also calculated. For the helical ribbon impeller, m = 0.304 Pa s^n whereas a value of m = 0.302 was obtained for the cone-plate geometry. Similar agreement was found for the rest of the single-phase fluids studied here. It is obvious that if both n and m are quite close for both geometries then the estimated viscosities are quite similar. These results can be explained by considering that a helical ribbon impeller produces a flow pattern which gives power input results similar to those predicted by a modified Couette flow analogy (8,9). The representation of the experimental power input data through Equation 6, is a clear indication that the constant K_s (see Equation 4) is a function of the fluid rheological properties (explicitly of n). These results are in agreement with those obtained at larger mixing scales with a similar impeller geometry (8). Then, the results of this and previous studies (5, 7- 9) clearly suggest that the helical ribbon is a geometry that can be used to estimate viscous properties of two-phase systems.

Typical experimental results of T as a function of N for two-phase fluids are shown in Figures 2 and 3 for three geometries (i.e. cone-plate, concentric cylinders and helical ribbon). As these figures suggest, shear thinning and shear thickening characteristics are observed for these types of fluids dependent on the value of N. This double behavior is determined through the value of n (exponent of N in Equation 3). For n < 1, the fluid exhibits shear thinning behavior whereas for n > 1, it is shear thickening. A similar behavior was observed for all fluids formulated with kaolin. It is also clear that as the kaolin concentration increases, the fluid becomes more shear thickening.

For the standard geometries (i.e. cone-plate and Couette), a more careful analysis of the data shown in Figure 3 (and in general for kaolin concentrations greater than 55%), shows that on increasing N the raw torque signal becomes quite unstable. This is particularly more severe for the cone-plate results. The instability of the torque signal even for long periods is a clear indication of particle-wall interactions and phase separation, which produce erroneous results regarding viscosity.

Transformation of the experimental raw data shown in Figures 2 and 3 into steady shear viscosity values implies the validity of the function K_p (n) given by Equation (6) for two-phase fluids. In other words, the distribution of shear rates (see Equation 4) cannot be obtained explicitly from raw torque-impeller rotational speed data without further assumptions because both n and m (from the power law model) are lumped in the A(n) function of Equation 3 and cannot be separated. Thus, for two-phase fluids and assuming that the Newtonian and single-phase non-Newtonian calibration is valid for two-phase fluids, the measured impeller rotational speed and the resulting torque were transformed into process viscosity by:

$$\eta_e = \frac{m\,K_p(n)\,N^{n-1}}{K_p} \tag{7}$$

In Figure 4, process viscosity, η_e, results for 60% kaolin are shown as an example of the typical results obtained for the two-phase fluids studied. The viscosity results are represented as a function of the impeller rotational speed and are compared with those obtained with classical geometries. This representation was chosen in order to

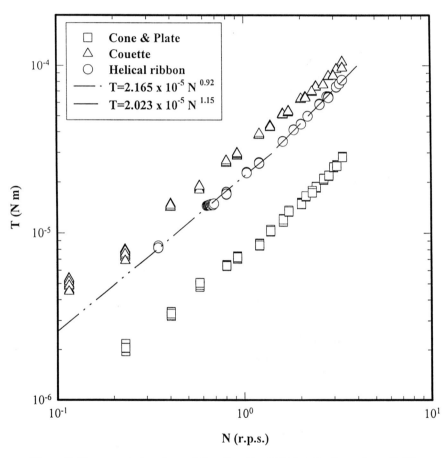

Figure 2. Torque-rotational speed function for 40 % kaolin (two-phase fluid). Comparison with results obtained with different geometries.

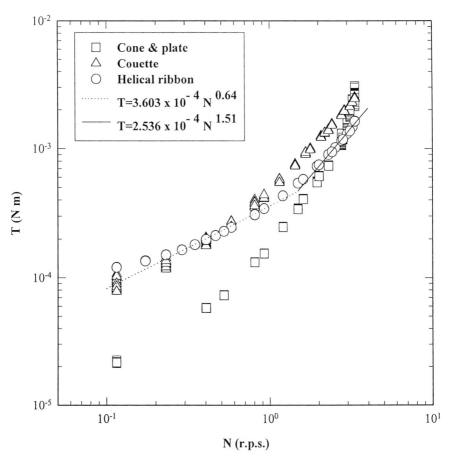

Figure 3. Torque-rotational speed function for 60 % kaolin (two-phase fluid). Comparison with results obtained with different geometries and predictions of Equation 8.

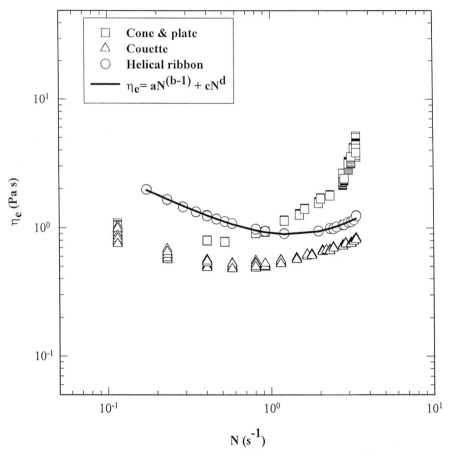

Figure 4. Process viscosity function for 60 % kaolin. Comparison with results from different geometries.

avoid more assumptions regarding the estimation of shear rate distributions inside the vessel. We believe this representation gives valuable information regarding the viscous response of the studied fluids. It is important to mention here that a model given by:

$$\eta_e = aN^{b-1} + cN^d \qquad (8)$$

was used to represent the experimental results for the two-phase systems. In Equation 8, a, b, c and d are fitting constants. The above model describes both the shear thinning and shear thickening characteristics of the two-phase fluids studied here. As the results of Figure 4 suggest, the helical ribbon impeller gives more stable viscosity values when compared with cone-plate and Couette geometries. In addition, Equation 8 represents quite well the experimental results.

Finally, all the above results were tested with a real two-phase fluid food. In Figure 5, the classical steady state flow curve obtained with the cone-plate geometry is shown for two samples of the same fluid. Although all the precautions for these tests were taken, it was impossible to obtain reproducible results. The variability in the results is quite important and it is obvious that no significant results regarding the viscosity of this fluid can be obtained with the cone-plate geometry. Then, the helical ribbon impeller was tested. In Figure 6, process viscosity results for the Mexican green sauce are presented for two different samples. This fluid clearly shows a shear-thinning behavior typical of this class of fluid foods. The shear thinning part of the model given by Equation 8 was successfully used to represent the experimental results. As this figure suggests, with this impeller geometry it is possible to obtain reproducibility.

Conclusions

Raw power consumption data (i.e. torque-impeller rotational speed) obtained with a helical ribbon impeller can be related to more fundamental rheological information by performing a careful calibration with single phase Newtonian and non-Newtonian fluids of known rheological properties. Once the mixer system is calibrated, it can be used to estimate viscous properties of more complex fluids (i.e. two-phase fluids). Estimation of absolute viscosity values using the mixer-viscometry principles requires further assumptions regarding the distribution of shear rates prevailing in the mixing system. The results of this study indicate that average shear rates are related to impeller rotational speed through a complex function of the flow behavior index. However, in order to express raw variables into process viscosities, this complex function is not necessary. The process viscosity results may be expressed in terms of more practical variables such as the impeller rotational speed.

The data obtained with the helical ribbon impeller could be more precisely related to food processing operations, particularly those related with formulations in agitated tanks. For two-phase fluid foods in which the particle-wall interactions are important, the flow curves, and thus viscosity curves, are highly inaccurate. In these

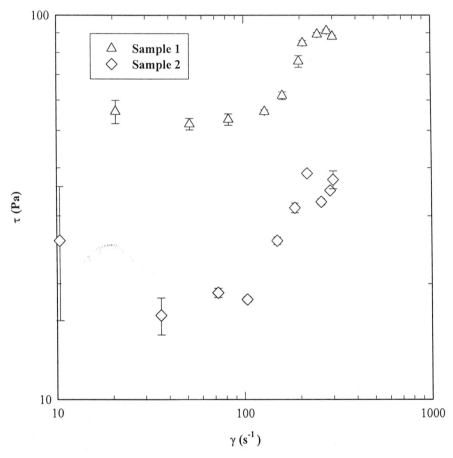

Figure 5. Flow curve for the Mexican green sauce obtained with a cone-plate geometry. Instabilities generated by particle-wall interactions.

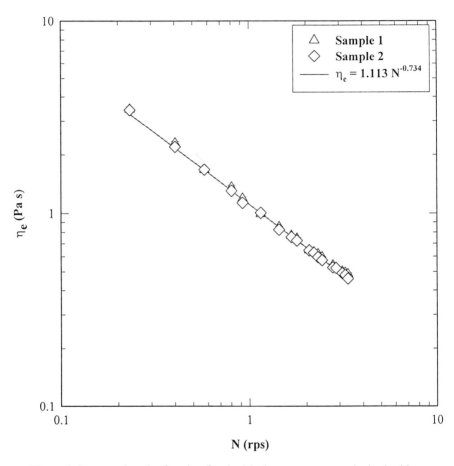

Figure 6. Process viscosity function for the Mexican green sauce obtained with a helical ribbon impeller.

cases, the mixer viscometry based in a helical ribbon is an alternative to estimate viscous properties.

Acknowledgments

The authors acknowledge the financial support of DGAPA-UNAM and CONACyT, México.

Literature Cited

(1) Barnes, H; Edwards, M.F.; Woodcock, L.V. *J. Chem. Eng. Sci.* **1987**, *42* (4), pp 591-608.
(2) Klein, B.; Laskowski, J.S.; Partridge, S.J. *J. Rheol.* **1995**, *39* (5) , pp 827-840.
(3) Castel-Perez, M. E., Steffe, J. M. In *Viscoelastic Properties of Foods*; Rao, M. A. and. Steffe, J. F., Eds.;Elsevier Applied Science: N.Y.; 1992; Chapter 10, pp 247-283.
(4) Tatterson, G.B. *Fluid Mixing and Gas Dispersion in Agitated Tanks;* McGraw-Hill Inc.: N.Y.; 1991, pp 325-411.
(5) Brito-De La Fuente, E.; Leuliet, J.C.; Choplin, L.; Tanguy, P.A. *A I Ch E Symp. Ser.* **1992**, *286* (88), pp 28-32.
(6) Metzner, A.B.; Otto, R.E. *A I Ch E J*, **1957**, *3* (1), pp 3-10.
(7) Doraiswamy, D.; Grenville, R.K.; Etchells, A, *Ind. Eng. Che Res.,* **1994**, *33*, pp 2253-2258.
(8) Brito-De La Fuente, E.; Choplin, L.; Tanguy, P.A. *Trans. I. Chem. E.*, **1997**, *75* (A), pp 45-52.
(9) Brito-De La Fuente, E.; Nava, J.A.; Medina, L.; Ascanio, G.; Tanguy, P.A. In *Proc. XIIth. Int. Congr. on Rheology;* Ait-Kadi, A, Dealy, J. M., James, D. F., and Williams, M. C. Eds.; Canadian Rheology Group: Quebec City, Canada 1996, pp 672-673.

PLANT PROTEIN FUNCTIONALITIES

Chapter 5

Production of High-Protein Flours as Milk Substitutes

S. H. Guzmán-Maldonado and O. Paredes-López

CINVESTAV-IPN, Unidad Irapuato. Apdo. Postal 629, 36500, Irapuato, México

The world has experienced a resurgence of interest in grain amaranth because it is one of the few species with potential to become an important source of dietary protein. The proteins of this grain have a better balance of essential amino acids than that of common grains; different studies indicate that amaranth protein is notably high in lysine. The Aztecs credited amaranth seeds with medical properties; recently, scientific information has accumulated supporting their nutraceutical characteristics. The feasibility of producing a high-protein flour (HPAF) was explored by us; two processes, liquefaction and saccharification were tested. As a result, a HPAF-L from liquefaction with 35.0% protein and a HPAF-S from saccharification (HPAF-S) with 29.4% protein were obtained. Essential amino acid profiles are roughly comparable to that of dry milk. Iron contents of both flours were higher than that of dry milk. Protein digestibility of HPAF-L and HPAF-S were 85 and 89%, respectively.

Amaranth Grain

The potential of amaranth grain as food was recognized by old cultures in America (*1*). In fact, in pre-Columbian times this grain was one of the basic foods of the New World, nearly as important as corn and beans. Thousands of hectares of Aztec, Inca, and other farmland were planted to the tall, leafy, reddish plants. Some 20,000 tons of amaranth grain were sent from 17 provinces to Tenochtitlán (present-day México city) in annual tribute to the Aztec emperor Moctezuma (*2*). For reasons not well-known, after the Spanish conquest, amaranth grain was discontinued as food by the Indians. Apparently, the use of this pseudocereal in pagan rituals and human sacrifice and with the collapse of Indian cultures following the conquest, caused amaranth to fall into disuse (*2*).

There are three species of grain amaranth: *Amaranthus hypochondriacus, A. cruentus,* and *A. caudatus*. These species are characterized by a profuse branching and a small terminal inflorescence (*3*). All three are still cultivated on a small scale in isolated valleys of Mexico, and Central and South America, where generations of farmers have continued to cultivate the crop of their forebears (*2*). Amaranth is a beautiful crop with brilliantly colored leaves, stems, and flowers of purple, orange, red, and gold. The seedheads resemble those of sorghum. The seeds, although barely bigger than a mustard seed (0.9-1.7 mm in diameter), occur in massive number, sometimes more than 50,000 to a plant (*3*).

For some years now, amaranth has been rediscovered as a useful and promising plant, mainly because it looks like an interesting alternative to alleviate the increasing need for food by some countries of the Third World. With a protein content of about 13 to 19%, amaranth seed compares well with the conventional varieties of wheat (12-14%), rice (7-10%), maize (9-10%), and other widely consumed cereals (*1, 2*). Moreover, amaranth seeds have a high level of the essential amino acid lysine, in comparison to the cited cereals; it contains 1.5 to 3.0 times more oil than other grains and can successfully grow in adverse environmental conditions and high saline soils. Amaranth is also a crop with multiple uses such as food and animal feed and has a promising potential in industrial applications, from cosmetics to biodegradable plastics (*1, 2, 4*).

Commercialization of amaranth has proceeded slowly. Numerous national and international meetings have been convened to report progress on the crop. Domestically some small foods companies began to sell amaranth as a health food, while pioneer growers and researchers formed organizations to advance the crop. However, consumer and market awareness was low while production capability outpaced demand (*5*).

Chemical Composition. A comparison between the average chemical composition of *Amaranthus* species with corn, whole rice, and wheat is shown in Table I. Amaranth flour is characterized by a relatively higher concentration of protein (13.0-19.0%) than in cereals (9.6 to 16%). The lipid content of amaranth grain is also relatively high (6.5-12.5%) depending on amaranth species, as compared to corn (5.3%), whole rice (2.4%), or wheat (2.4%), and starch content (56-78%) is close to the most important cereals (76-85%) (*6, 7*).

Amino Acid Content. The amaranth amino acid composition is unusual because its essential amino acid balance is closer to the optimum required in the human diet (Table I) (*7-9*). Different studies indicate that amaranth protein is notably adequate in lysine (4.8 to 6.4 g/100g protein) as compared with the FAO/WHO/UNU reference patterns (*6, 7*). In contrast, cereal grains are deficient in lysine; corn is also deficient in tryptophan, and rice and wheat proteins are limiting in lysine and threonine (*6*). Besides, the sulfur amino acid concentration (3.7 to 5.5 g/100g protein) of amaranth is higher than that of the most important legumes (1.4 g/100g protein) (*7*). These characteristics make this promising crop an excellent source of dietary proteins (*9*). Isoleucine, leucine, and valine, other limiting amino acids in some foods, are not considered a serious problem since they are found in excess in most common grains (*7*).

Amaranth Protein
Protein Concentrates and Isolates. It has been reported that the greater part of amaranth grain protein is found in the embryo, a ring surrounding a diploid starchy perisperm (*6*). Bressani (*10*) suggested that such distribution is responsible for the high protein content. This characteristic enabled Sánchez-Marroquín *et al.* (*11*) to produce a protein concentrate by increasing the protein content of amaranth seed flour from 18 to about 33%. The process involves mostly milling, air drying and air separation giving rise to two flour streams: one rich in protein and another rich in starch. The cited authors claim that this air classification process may increase protein content of the original flour by three- to four-fold.

Table I. Chemical Composition and Essential Amino Acid Content of Seed from Amaranth and Some Cereals

Component	Amaranth	Corn	Whole rice	Wheat	FAO/WHO/UNU Adult	FAO/WHO/UNU Child
Chemical composition (%, dry basis)						
Crude protein[1]	13.0-19.0	11.9	9.6	16.0		
Crude fat	6.5-12.5	5.3	2.4	2.4		
Crude fiber	3.9-17.5	2.7	1.0	3.0		
Ash	3.2- 3.9	1.6	1.6	2.2		
Starch	56.0-78.0	78.5	85.4	76.4		
Amino acid (g/100g protein)						
Histidine	2.4- 3.2	3.0	--	1.9	1.6	1.9
Isoleucine	3.5- 4.1	3.5	4.7	3.1	1.3	2.8
Leucine	5.0- 6.3	12.4	8.5	6.6	1.9	9.3
Lysine	4.8- 6.4	3.0	4.0	1.9	1.6	6.6
Methionine + cysteine	3.7- 5.5	4.3	4.4	4.1	1.7	4.2
Phenylalanine + tyrosine	7.1- 9.1	7.7	10.3	7.6	1.9	7.2
Threonine	3.3- 4.6	3.3	3.8	2.4	0.9	4.4
Tryptophan	1.0- 4.0	0.7	1.2	1.5	0.5	1.7
Valine	3.2- 4.8	4.9	7.0	3.4	1.3	5.5

[1]Protein conversion factors (factor x %N) were: amaranth, 5.85; corn, 6.25; rice, 6.25; wheat, 5.7

Adapted from references *6-9*, and *10*.

Efforts also have been made to obtain protein isolates from amaranth flour. As a result, amaranth protein has been extracted with sodium hydroxide solutions at various pHs. The recovery level varied from 71-74 g/100g of soluble protein. The isolates contained 81% protein. Paredes-López *et al.* (*12*) obtained extracts of salt-soluble proteins with 1.0 M sodium chloride, giving protein recoveries of nearly 51%. Moreover, different studies deal with the characterization of the salt-soluble amaranth globulins (*13, 14*), which are the most important proteins from a nutritional point of view. The quantities of albumins plus globulins were found to vary between 39.1 and 82%; albumins may represent from half up to two-thirds of these amounts. The alcohol-soluble prolamins varied from 1.2 to 7.2%, and glutelins from 7.0 to 46.5% (*12, 15*). Barba de la Rosa *et al.* (*14*) have found that globulins are composed of 10S and 12.7S fractions with subunits of molecular weights in the range of 23-52 kDa. More recently, Sanfeng and Paredes-López (*16*) extracted and characterized an 11.9S amaranth globulin. The variations in sedimentation coefficients may be due to different extraction procedures.

Molecular Biology of Amaranth Proteins. Molecular biology techniques are the most powerful tool for transfer of specific and new genetic information into living organisms. Currently, the procedures of plant molecular biology are already established for many species and are used for research and applied purposes (*1*). Through biotechnology and molecular biology, plants, as well as other living organisms can be used either as receptors of genes introduced by genetic transformation or as a source of interesting genes to isolate, characterize, and engineer for their introduction into the same plant or into different plants or organisms.

Three concurrently steps are usually carried out in gene isolation: 1) obtention of a cDNA library from mRNA; 2) production of a DNA probe based on amino acid sequence of a protein of interest and, 3) obtention of a genomic library from total DNA (Figure 1). The goal is to identify both the original gene and corresponding promoter in order to combine them. Once the gene of interest is isolated, it can be used for transforming other plants or to express in the same amaranth plant. Following this scheme, Barba de la Rosa *et al.* (*17*) extracted and purified a globulin from amaranth by gel filtration and ultracentrifugation; this protein was called amarantin (*16*). The protein, with molecular weight around 398 kDa, gave by SDS-PAGE, under reductive conditions, polypeptides of 59, 36 and 24 kDa. The N-terminal of the 59 kDa polypeptide was sequenced. Also, a cDNA library was obtained from poly(A+)RNA isolated from developing amaranth seeds; as a result, amaranth globulin cDNA clones were identified by hybridization with a synthetic oligonucleotide designed from the N-terminal amino acid sequence. Eight globulin clones were distinguished and two of them were sequenced, resulting in identical sequences. One of the clones was subcloned in pSPORT1. Currently, Osuna and Paredes-López (*18*) are working to express the polypeptide in transformed *E. coli* with a pT7-AMAR plasmid which contains an IPTG induced promoter.

Production of High-Protein Amaranth Flours
Amylolytic Enzymes. There are essentially five groups of enzymes involved in the hydrolysis of starch (*19*). The endo- and exoenzymes act primarily on the α-1,4 linkages (e.g. α-amylase, β-amylase and glucoamylase), whereas the debranching enzymes act exclusively on the α-1,6 linkages (e.g. pullulanase, isoamylase). A fourth

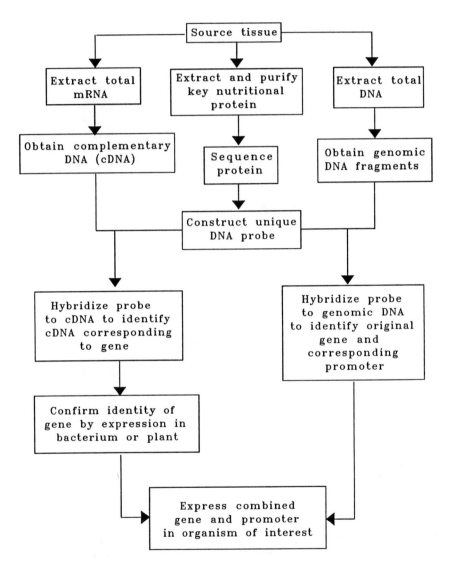

Figure 1. Basic steps on gene isolation.

group, the isomerases (e.g. glucose isomerase), act on glucose to convert it to fructose. Finally, cyclodextrin glycosyltransferase degrades starch by catalyzing cyclization reactions. Traditionally, starch transformation industries have used acid processing to obtain most of their products (*19*). But, amylolytic enzymes have gained importance because of their ability to catalyze batch reactions under moderate conditions, which are not reached with acid processes wherein nonspecific hydrolysis and formation of byproducts are present.

Enzyme production is carried out based on the optimal conditions of growth for the microorganism being used. Thus, pH, temperature, aeration and control of contaminants, are some of the parameters which have to be controlled (*20*). Most of the enzyme production is carried out using liquid fermentation; however, solid state fermentation techniques have also been employed for the production of crude enzyme preparations for the food industry (*21*). Molecular biology techniques opened a new framework for amylolytic enzyme production. Steyn and Pretorius (*22*) described the co-expression of *Sacharomyces cerevisiae* var. *diastaticus* glucoamylase-encoding gene, STA2, and a *Bacillus amyloliquefaciens* α-amylase-encoding gene, AMY, in *S. cerevisiae*. As a result, co-expression of these two genes synergistically enhanced starch digestion; this combination may have potential in food industry applications for saccharification. Another attempt at combining enzyme activities was the experiment conducted by Shibuya *et al.* (*23*); an α-amylase/glucoamylase fusion gene which was successfully expressed in *S. cerevisiae*, was constructed. The resulting fusion protein was found to have higher raw-starch-digesting activity than those of the original α-amylase and glucoamylase. Protein engineering experiments also have been conducted to improve the properties of amylolytic enzymes for specific food industrial applications (*20*).

Liquefaction Procedure. The liquefaction procedure was developed as a basic procedure for enzymatic hydrolysis of amaranth flour to produce a high-protein amaranth flour (HPAF) and a by-product called maltodextrins. Liquefaction is the term given to a procedure in which α-amylase is used in order to hydrolyze amaranth starch to obtain a dextrose equivalent (DE, degree of hydrolysis) from 5 to 19 (*19, 24, 25*). Figure 2 shows the process developed for the liquefaction of the amaranth.

Slurries containing 20g/100mL raw flour were prepared with distilled water, adjusting to pH 6.5, and gelatinized in boiling water for 10 min. Sample flasks were equilibrated in a shaking bath at 70 °C, then 0.01% (v/w) α-amylase was added and kept at these conditions for 5 min. The enzyme concentration had a total activity of 3790 MWU (Wohlgemulth units/g enzyme; with one MWU being defined as that activity which will dextrinize 1 mg of soluble starch to a defined blue value in 30 min) (*25*). After hydrolysis, the samples were cooled rapidly in ice water and centrifuged at 11,000g for 30 min at 4 °C. The precipitate, HPAF, was freeze-dried and kept at 4 °C until analysis. The enzyme present in the supernatant was inactivated by adjusting the pH to 3.5-4.5 and placing the sample flasks in boiling water for 10 min. The supernatant was also decolorized with 2.5 g/100mL of activated charcoal at 60 °C for 30 min (*25*). Supernatant, containing the maltodextrins, was freeze-dried and kept at 4 °C until analysis.

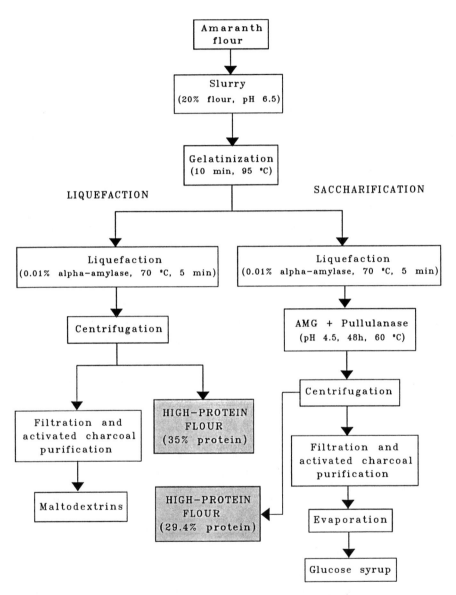

Figure 2. Basic procedure to produce high-protein flours.
AMG = glucoamylase

Saccharification Procedure. This process involved two steps: 1) amaranth flour liquefaction with α-amylase in order to obtain a DE<12.0, and 2) saccharification of liquefied flour to obtain HPAF and the by-product, a glucose syrup (Figure 2). Saccharification was carried out by the amylolytic enzymes glucoamylase (AMG) and pullulanase. The procedure was as follows: the liquefied samples were adjusted to pH 4.5 and equilibrated at 60 °C; then 0.50 mL/100g flour of AMG was added to the samples, plus a fixed amount of pullulanase (30 µL/100g flour). The 1, 6-α-glucosidic linkages act as a barrier to the rapid action of the exo-acting saccharifying amylase AMG, which slowly hydrolyzes such linkage in amylopectin. When a debranching enzyme such as pullulanase is used simultaneously with AMG to saccharify a partially hydrolysed starch, an increase in glucose yield is obtained (*24, 25*). For this purpose, we decided to use the pullulanase dosage recommended by the supplier. The glucoamylase concentration had a total activity of 10 DU (Diazyme unit, with one DU being defined as the amount of enzyme necessarily to liberate 1g of reducing sugars as glucose per h). After saccharification, the samples were cooled rapidly in ice water and centrifuged at 11,000g for 30 min at 4 °C. Precipitate, as HPAF, was freeze-dried and kept at 4 °C until analysis. The supernatant glucose syrup was decolorized as in the liquefaction process. The glucose solution was finally concentrated to 70% (dry basis) with a Rotaevapor.

Chemical Composition of High-Protein Flours and Byproducts

High-Protein Flours. Table II shows the increase in protein content when whole amaranth flour (15.6% protein) was liquefied with α-amylase in order to produce high-protein amaranth flour (35.0% protein) (HPAF-L). As a result of the enzymatic digestion, the starch content decreased to approximately 50%. Fat, crude fibre, ash and free sugars were enhanced at different levels.

Protein content of high-protein flour obtained from the saccharification procedure (HPAF-S) was lower (29.4% protein) than that of flour obtained from liquefaction. The lower protein content might be due to the prolonged time of saccharification (48 h), which causes an increase in solubilization of protein. As can be seen in Table II, protein contents of both flours were higher than that of dry milk (27.6%). Free sugars of HPAF-S were increased notably (19.8%); starch decreased from 66.4% to 23%.

Byproduct Quality

Maltodextrins. Chemical analysis showed that protein, fat and ash remained mostly in the high-protein flour (data not shown here). Protein content of amaranth maltodextrin was about 0.9% (dry basis). It is important that protein content of carbohydrate products (e.g. maltodextrins, glucose syrup) be reduced to levels lower that 1% to prevent amino acids or small peptides reacting with sugars to form colored complex compounds (*25*). Total color differences (ΔE), in relation to the white standard, of amaranth maltodextrins was commercially aceptable. After 90 days of storage, this product showed a minor increase in ΔE value; in other words, amaranth maltodextrins developed a visually undetectable darker color.

Glucose Syrup. In spite of having subjected amaranth glucose to activated charcoal

Table II. Composition of High-Protein Amaranth Flours, Dry Milk and Yield

Component (% dry basis)	Whole amaranth flour	High-protein flour		Dry whole milk
		Liquefaction	Saccharification	
Protein	15.6	35.0	29.4	27.6
Fat	7.3	13.1	15.5	26.0
Fiber	5.2	8.7	10.8	---
Ash	3.0	2.9	2.9	6.1
Soluble sugars	2.5	7.0	19.8	39.0
Starch	66.4	33.3	23.0	---
Yield	100.0	49-52	40-42	---

treatment for clarification and protein adsorption, protein content was on the order of 1% (dry basis); the protein content of commercial glucose is only 0.1%. This higher glucose consentration causes development of a pale yellow color in amaranth glucose after 30 days of storage, which is commercially acceptable. Other characteristics, such as total solids, ash, sulfur dioxide and pH of amaranth glucose (data not shown here), were similar to those of commercial glucose.

Nutritional and Functional Characteristics of HPAF
Nutritional Characteristics. As a result of enzymatic hydrolysis, the essential amino acid lysine increased from 6.2 g/100 g protein (raw flour) to 7.2 g/100 g protein for HPAF-L and 7.7 g/100 g protein for HPAF-S (Table III). In contrast, the sulfur amino acids were slightly lower that those of raw flour, but almost the same as that of dry milk. The other essential amino acids were comparable to those of dry whole milk.

In vitro protein digestibility increased from 74.0% (raw flour) to 85.2 (HPAF-L) and 88.7 % (HPAF-S) (data not shown here). It is well known that cooking amaranth grain increases the protein efficiency ratio (PER) from 1.7 (raw amaranth) to 3.2, the net protein utilization (NPU) from 73.9% to 76.8% and the biological value (BV) from 73.0% to 86.0% (*10*). Similar determinations are pending for HPAF-L and HPAF-S.

High levels of Ca, Fe, and Mg were obtained in both high-protein flours after liquefaction and saccharification processes (Table III). Calcium content of HPAF-L increased from 1.5 mg/g sample (raw flour) to 3.9 mg/g sample; therefore, a 100 g portion of this flour provides 39% of calcium of the Reference Daily Intake (RDI) issued by the U.S. Food and Drug Administration and the U.S. Department of Agriculture (*26*). Iron increased from 0.154 mg/g sample (raw flour) to 0.184 mg/g sample of HPAF-L. A 100 g portion of this flour could provide 123% of the iron requirement according to the RDI. Iron content was significantly higher in both flours compared to that of dry milk. Calcium and iron contents of HPAF-S were slightly lower than those of HPAF-L.

Functional Characteristics. Enzymatic hydrolysis significantly increased flour dispersibility for both HPAF-L (75%) and HPAF-S (94%) as compared to that of raw whole flour (27%) (data not shown here). The water holding capacity was also higher in both high-protein flours (2.5 mL H_2O/g) than that of raw flour (1.7 mL H_2O/g). Total color differences (ΔE) was lower for both products compared to dry whole milk (Figure 3).

In conclusion, chemical composition, and nutritional and functional characteristics of HPAF-L and HPAF-S are roughly comparable to those of dry whole milk. Therefore, these flours can be used as whole milk extenders or as milk substitutes.

Potential Nutraceutical Properties of Amaranth Flours
In addition to the nutrients that are involved in normal metabolic activity, plant foods contain components that may provide additional health benefits. These food components, which are derived from naturally occurring ingredients, are generally referred to as nutraceuticals (*6*). In recent years, the number of nutraceutical or

**Table III. Amino Acids and Mineral Contents of High-Protein Amaranth
Flours and Dry Whole Milk**

Component	Whole flour	HPAF-L	HPAF-S	Dry Whole milk	RDI (mg) (1994)
Amino acid (g/100g protein)					
Essential					
Histidine	2.8	4.5	4.1	-----	
Isoleucine	4.4	3.8	3.2	7.5	
Leucine	7.2	7.5	7.4	11.9	
Lysine	6.2	7.2	7.7	8.7	
Methionine +					
Cysteine	4.8	4.5	3.5	4.2	
Phenylalanine +					
Tyrosine	6.2	7.4	6.9	11.5	
Threonine	3.5	3.7	4.0	4.7	
Tryptophan	ND	ND	ND	1.5	
Valine	4.8	4.5	5.0	7.0	
Nonessential					
Alanine	4.0	4.4	3.5		
Arginine	8.4	7.9	7.6		
Aspartic acid	8.4	8.1	7.3		
Glutamic acid	15.7	14.6	12.7		
Glycine	7.3	7.8	7.6		
Proline	4.7	4.3	4.2		
Serine	6.6	6.1	6.7		
Mineral (mg/g sample)					
Calcium	1.5	3.9	2.9	12.4	1000
Iron	0.154	0.184	0.174	0.003	15
Magnesium	3.3	4.3	3.5	------	400
Potassium	5.2	3.7	2.7	17.11	

HPAF = high-protein amaranth flour; L = liquefaction; S = saccharification
RDI = reference daily intake (26)
ND = not determined

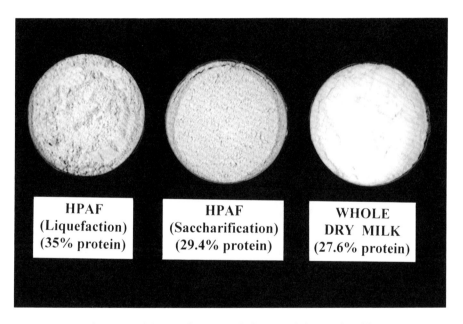

Figure 3. High-protein amaranth flours and dry whole milk.

functional foods that have potential benefits for health has grown tremendously. One health food niche for amaranth protein flours is as a gluten- or prolamin-free protein. Celiac disease is characterized by sensitivity to the prolamin fraction of cereals. Prolamins are present in amaranth in the range of 0.7-11.0% depending on specific extraction method and variety (5, 7). Another nutraceutical aspect of amaranth proteins is their globulins subunits 7S and 11S; protein isolation shows that amaranth globulins and amarantin, the 11S fraction of globulins, represents 20.5 and 18.6% of the total seed protein, respectively (14). It has been suggested that soy globulins 7S and 11S exert a cholesterol-lowering effect (27). Moreover, due to the outstanding protein and iron contents of amaranth, protein concentrates obtained from this grain have been promoted as bodybuilding foods (5).

Amaranth oil contains a high degree of unsaturation with about 53-95% linoleic and oleic, 0.3-1.3% linolenic, and 2.2-5.4% stearic acid based on total oil (15). It is well established that essential fatty acid deficiencies result in lymphoid atrophy and depressed antibody responses. Small dietary amounts of linoleic acid are required for the normal propagation and maturation of a cell-mediated immune response (6). It is important that HPAF obtained from liquefaction and saccharification increased their lipid content from 7.3% (amaranth whole flour) to 13.1 and 15.5%, respectively, which simultaneously increases the potential nutraceutical effects of such essential fatty acids.

Two well-established problem minerals in human health are calcium and iron (6). Deficiencies in these minerals are relatively small among people in developed countries because of the greater use of animal food groups (26). On the other hand, mineral deficiencies are common in developing countries where plant foods are the major nutrient sources. Dietary calcium deficiencies have been linked epidemiologically to several chronic diseases including, osteoporosis, hypertension, and colon cancer. Iron deficiency, on the other hand, is characterized by decreases in red blood cell function, as well as in depressed function of numerous other cellular activities because of inadequate oxygen delivery (6).

In developing countries, animal proteins are either too expensive (Latin America and Africa) or not readily accepted (India). Daily intakes of milk or milk products are one of the lower intakes in such countries; because of this, new sources of good quality proteins are very important.

Literature Cited

1. Lozoya-Gloria, E. In *Amaranth- Biology, Chemistry and Technology*; Paredes-López, O., ed.; CRC Press: Boca Raton, FL, 1994, Chapter 1, pp 1-7.
2. National Research Council. *Amaranth- Modern Prospects for an Ancient Crop;* National Academy Press: Washington, DC, 1984, Chapter 1, pp 1-12.
3. Espitia-Rangel, E. In *Amaranth- Biology, Chemistry and Technology*; Paredes-López, O., Ed.; CRC Press: Boca Raton, FL, 1994, Chapter 3, pp 23-38.
4. Schnetzler, K. A.; Breene, W. M. In *Amaranth- Biology, Chemistry and Technology*; Paredes-López, O., Ed.; CRC Press: Boca Raton, FL, 1994, Chapter 9, pp 155-183.
5. Lehemann, J. W. *Cereal Foods World.* **1996,** *41(5),*399-411.
6. Guzmán-Maldonado, S. H.; Paredes-López, O. In *Processing and Evaluation of Functional Foods*; Mazza, G. J., Ed.; Technomic Publishing Co.: Inc. Lancaster, PA, 1998. In press.

7. Segura-Nieto, M.; Barba de la Rosa, A. P.; Paredes-López, O. In *Amaranth-Biology, Chemistry and Technology*; Paredes-López, O., Ed.; CRC Press: Boca Raton, FL, 1994, Chapter 5, pp 75-105.

8. Wrigley, C. W. In *Wheat- Chemistry and Technology*; Pomeranz, Y., Ed.; American Association of Cereal Chemist. Inc.: MN, 1988, Vol I, Chapter 5, pp 159-173.

9. Luh, B. S.; Liu, Y-K. In *Rice: Production & Utilization*; Luh, B. S., Ed.; AVI Publishing Company, Inc.: Westport, CT, 1980, Chapter 17, pp 590-613.

10. Bressani, R. *Food Rev. Int.* **1989,** *5,*13-38.

11. Sánchez-Marroquín, A.; Domingo, M. V.; Maya, S.; Saldaña, C. *J. Food Sci.* **1985,** *50,*789-794.

12. Paredes-López, O.; Mora-Escobedo, R.; Ordorica-Falomir, C. *Lebensm. Wiss. u Technol.* **1988,** *21,*59-61.

13. Konishi, Y.; Fumita, Y.; Ikeda, K. *Agric. Biol. Chem.* **1985,** *49,*1453-1459.

14. Barba de la Rosa, A. P.; Paredes-López, O.; Gueguen, J. *J. Agric. Food Chem.* **1992,** *40,*937-942.

15. Paredes-López, O.; Guzmán-Maldonado, S. H.; Ordorica-Falomir, C. In *New and Developing Sources of Food Proteins*; Hudson, B. J. F., Ed.; Chapman Hall: London, 1994, Chapter 8, pp 241-279.

16. Sanfeng, CH.; Paredes-López, O. *J. Food Biochem.* **1997,** *21,*53-65.

17. Barba de la Rosa, A. P.; Herrera-Estrella, A.; Utsumi, S.; Paredes-López, O. *J. Plant Phys.* **1996,** *149,*527-532.

18. Osuna, C. J.; Paredes-López, O. Personal communication, **1997.**

19. Guzmán-Maldonado, H.; Paredes-López, O. *Critical Reviews in Food Science and Nutrition* **1995,** *35(5),*373-403.

20. Jennylynd, A. J.; Byong, H. L. *J. Food Biochem.* **1997,** *21,*1-52.

21. Paredes-López, O.; Guzmán-Maldonado, S. H.; Alpuche-Solís, A. In *Bioconversion of Waste Materials to Industrial Products;* Cross, C., Ed.; Chapman & Hall: U.K. 1997, Chapter 3, pp 103-153. In press.

22. Steyn, A. J. C.; Pretorius, I. S. *Gene* **1991,** *100,*85-93.

23. Shibuya, I.; Tamura, G.; Shima, H.; Ishikawa, T.; Ahara, S. *Biosci. Biotechnol. Biochem.* **1992,** *56(6),*884-889.

24. Guzmán-Maldonado, H.; Paredes-López, O.; Domínguez, J. *Lebensm. Wiss. u. Technol.* **1993,** *26,*28-33.

25. Guzmán-Maldonado, H.; Paredes-López, O. *Process Biochem.* **1994,** *29,*289-293.

26. Anderson, J. J. B.; Allen, J. C. In *Functional Foods, Designer Foods, Pharmafoods, Nutraceuticals*; Golberg, I., Ed.; Chapman & Hall: London, 1994, Chapter 15, pp 323-338.

27. Kitamura, K. *Second International Symposium on the Role of Soy in Preventing and Treating Chronic Diseases.* Sept. 15-18, 1996, Brussels, Belgium. pp 48.

Chapter 6

Functional Properties of Soy Proteins

N. S. Hettiarachchy and U. Kalapathy

Department of Food Science, University of Arkansas, Fayetteville, AR 72704

Soy proteins, due to their structural characteristics, deliver a range of functionalities in food products. Extrinsic factors such as temperature, pH, ionic strength, and interaction with other components can also affect protein functionality. Hence, by controlled manipulation of processing conditions, functionalities can be improved to a certain extent. However, further improvement may require changes in protein structure. Chemical, and enzymatic methods can be used to obtain desirable properties by structural modification of soy proteins. Chemical modifications such as phosphorylation, deamidation, acylation and succinylation have been used to improve emulsifying and foaming properties of soy proteins. Hydrolysis of soy protein using enzymes such as trypsin, pepsin, and papain, and protein cross-linking by transglutaminase have been used to improve the functionalities. Adhesive, emulsifying and foaming properties of soy proteins can be utilized in industrial products such as wood adhesive, cosmetic products, and packaging materials.

Soy proteins, due to their diverse physicochemical properties resulting from the structural characteristics, deliver a range of functionalities in food systems. In general, functionalities of proteins are defined as their physicochemical properties and behavior that impart desirable characteristics in a food system. Intrinsic factors such as protein size, hydrophilic/hydrophobic characteristics, and conformational structure, and extrinsic factors such as temperature, pH, ionic strength, and interaction between protein molecules as well as other food components, including water, lipids, and carbohydrates, play a major role in determining protein functionalities in a food (1).

80

Influence of Intrinsic Factors

Intrinsic factors affect the behavior of proteins in food systems during processing, production, and storage (2). Researchers have attempted to correlate the physicochemical properties of proteins with their functional properties in food systems. These correlations were based on the changes in selected physicochemical properties that are dictated by changes in protein functionalities (Table I) (2). Major components of soy proteins are 11S (about 52%) and 7S (about 35%). The average molecular weights of 7S and 11S proteins are 180 and 340 kDa (3). While 7S proteins have only 2-3 cysteine groups per mole, 11S proteins have 18-20 disulfide groups. As a result of the structural differences of 7S and 11S their functional properties vary depending on the influence of extrinsic factors such as pH, temperature, and ionic strength. The effect of these extrinsic factors in relation to structural differences of 7S and 11S proteins will be briefly discussed in the following section.

Table I. Correlation of Protein functionalities with Molecular Properties

Molecular Properties	Functional Correlation
Size	Larger size allows more interfacial interaction; stronger films
Amphiphilic properties	Distribution of polar and apolar residues; interfacial interaction; foaming, emulsion
Flexibility	Unfolding at interface; facilitate interfacial interaction; films, foaming, emulsion
Charged groups	Protein-protein interaction in films; repulsion between bubbles in foams; hydration and water solubility

Influence of Extrinsic Factors

Effect of thermal treatment of soy proteins on functional properties has been well documented (4). It has been shown that heat treatment during isoelectric precipitation of soy protein had desirable effects on functional properties (5). The protein isolate heated at 60 °C showed noteworthy flow characteristics, possessed the highest water-holding capacity and a good solubility (5). Water-binding capacity and viscosity of soy proteins have been shown to increase due to denaturation resulting from heat treatment at 80-100 °C for 30 min. Heat treatment also resulted in improved emulsifying properties due to increased hydrophobicity (6). Soy proteins form heat-induced gels at a concentration higher than 8 %. The heat-induced gels from 11S fraction showed higher gel strength and greater water holding capacity than those obtained from 7S fraction. These differences were due to the presence of disulfide bonds in the 11S proteins which influence the dissociation/association and unfolding behavior of protein subunits. Sulfide bonds in 11S proteins promote three dimensional matrix formation during gelation which results in stronger gels with greater water-holding capacity.

Alkali treatment also results in protein unfolding, disulfide cleavage and dissociation. High pH treatments often leads to more severe denaturation than those caused by thermal treatments (7). A combined treatment of soy proteins at about 65 °C and a pH of 10-11 resulted in an optimum exposure of hydrophobic groups to exhibit higher emulsifying capacity (8). The 11S fraction is more susceptible to alkali denaturation than the 7S fraction due to their high disulfide content (about 18-20 disulfide bonds per mole).

Ionic salts disrupt electrostatic interactions and in turn affect protein-protein and protein-solvent interactions. The changes in these interactions can lead to conformational changes and influence the functional properties. The solubility of soy proteins is significantly affected by ionic strength of the solution. This ionic strength effect is highly pH dependent. At pHs higher than 7.0, the solubility progressively decreases with increasing ionic strength. At pH 7.0, the ionic strength effect is insignificant. Although precise explanation is associated with the basic structural features, in general solvation, salting out and salting in properties are attributed for this behavior. High ionic (about 1 M) strength stabilizes the quaternary structure of soy proteins against dissociation and denaturation, and hence affects the reversible association-dissociation and unfolding properties.

Functional Properties

Water-Hydration Properties. The water as a solvent plays an important role in determining protein conformation, as well as the functionality of soy proteins. Soy proteins interact with water through their peptide bonds (dipole-dipole or hydrogen bonds) and amino acid side chains (ionized, polar, and nonpolar groups). Therefore, the characteristics of soy protein structures such as accessible surface polarity and hydrophobicity, as well as surface area and topography, are responsible for water sorption and hydration properties (9).

Water-Holding Capacity. The water holding-capacity of soy proteins plays a major role in the textural properties of various foods, especially comminuted meats and baked doughs. The amount of water associated with proteins depends to a large extent on the amino acid composition, protein conformation, surface hydrophobicity, pH, ionic strength, temperature and concentration (9). While water absorption and binding are important in food formulations, water resistance is an important adhesive property that determines the adhesive bond durability (10). Adhesives prepared from soy meal had poor water resistance due to the presence of carbohydrates.

Water Solubility. A high protein solubility is required to obtain desirable functionalities. Acidic conditions precipitate soy proteins while basic conditions usually improve the solubility of the proteins. Disulfide bond-reducing agents, such as thiols (cysteine, mercaptoethanol) and sulfites, by cleaving disulfide bonds cause disaggregation of proteins and enhance solubility (11). Improvement on soy protein solubility by enzyme (pepsin, papain, ficin, trypsin, bacterial and fungal proteases) hydrolysis has been reported (12, 13).

Matrix Forming Properties. Covalent protein-protein interactions (e.g. disulfide bonds between peptide chains and non-covalent protein-protein interactions (e.g., hydrophobic and van der Waals interactions, hydrogen bonding between two electronegative atoms, and ionic interactions between charged amino acid side chains) are important in functional properties of soy proteins such as gelation and film formation (14). The major soy protein fractions glycinin (11S fraction) and conglycinin (7S fraction) both have the ability to form ordered gel structures. These proteins have complex quaternary structures that easily undergo association-dissociation. Utsumi and Kinsella (15) investigated the forces involved in soy protein gelation and showed that electrostatic interactions and disulfide bonds were involved in the formation of 11S globulin gels, hydrogen bonding as involved in the formation of 7S globulin gels, and both hydrogen bonding and hydrophobic interactions were involved in soy isolate gels.

Gelation. Soy proteins can form gels and provide a structural matrix for holding water, flavors, sugars, and food ingredients in various food applications, and in new product development. Tofu gel (soft, hard, or dried) is a protein matrix which incorporates calcium, or magnesium, and also serve as a source of nutrients. Protein gelation is useful not only for the formation of solid viscoelastic gels, but also for improving water absorption, thickening, particle binding (adhesion), and emulsion or foam-stability. The effect of intrinsic structural properties and extrinsic factors such as pH, ionic strength on the soy protein gelation has been extensively investigated (15, 16). Protein concentration, temperature and duration of heating, salt concentration, thiols, sulfite and lipids affect gel formation and properties (17).

Film Formation. Concentrated soy protein solutions can be thermally coagulated on a flat container and dried to form films. Thin protein-lipid films are formed on the surface of soy milk by incubating for a few hours at 95 °C. Films can be prepared repeatedly by simply removing the previously formed film from the surface. Soy protein films can be used in edible packaging applications for food preservation and also as a carrier of nutritional and non-nutritional food additives. Edible films from biopolymers such as soy proteins have been extensively reviewed by Kester and Fennema (18), Krochta (19); and Gennadios et al, (20).

Surface Active Properties. Surface properties of proteins are primarily determined by their structural flexibility and amphiphilic character which govern the ability of proteins to form stable interfaces between two immiscible phases such as oil/water, and air/water. Hence, any interactions which influence the formation and stabilization of the interface will affect these surface properties. The hydrophobic interactions between apolar regions of soy proteins (alkyl side chains of amino acids such as alanine and leucine, or aryl side chain of phenylalanine) and apolar aliphatic chains of the lipid (hydrocarbon chain of fatty acids such as oleic acid) are important forces which determine emulsifying properties. Denaturation or partial hydrolysis unfolds soy proteins, resulting in a higher surface hydrophobicity and stronger interaction with lipids, and hence lead to improved emulsifying properties. High pressure homogenization also can increase the degree of protein-lipid interaction by increasing interfacial area.

Emulsion. Soy proteins aid in the formation of emulsions, mainly by decreasing interfacial tension between water and oil, and help stabilize the emulsion by forming a physical barrier at the interface. Surface hydrophobicity and solubility are the major factors that determine the emulsifying activity while the molecular flexibility of proteins is important for the emulsion stability (21). In native state, globular soy proteins with stable structure and lower surface hydrophilicity are poor emulsifying agents.
The ability of protein to aid formation and stabilization of emulsions is important for many food applications including coffee whiteners, mayonnaise, salad dressings, frozen desserts, and comminuted meats.
The emulsifying properties of soy proteins have been extensively studied (3, 22). Soy isolate shows greater emulsifying capacity compared to flours and concentrates. The pH, ionic strength, and temperature affect emulsification of soy proteins (22, 23). Studies indicate that emulsion properties were optimum at alkaline conditions. Poor emulsifying properties were obtained in the pH range of 5.0 to 6.0. Emulsion capacity decreased when temperature increased above 50 °C. The emulsifying capacity of soy increased by 42% when ionic strength of protein dispersion increased from 0.03 to 0.05 at pH 7.0.

Foaming. Foamability and foam stability are important functional properties of proteins essential in many food formulations. Many formulated foods are either foams, emulsions or gels and proteins are the preferred surface active and network-forming agents in these foods (24). Although soy proteins have good foaming capacities, practical applications of these proteins in foams are limited due to their instabilities. For foam formation proteins should be water soluble, and flexible to form cohesive film at the air water interface. In order for the films to be stable, they should possess sufficient mechanical strength and viscosity for preventing subsequent rupture and coalescence.
Soy proteins exhibit foaming properties, and studies revealed that soy isolates are superior to soy flours and concentrates (25, 26). Solubility of soy protein is closely correlated to foaming properties. Lipid materials are responsible for destabilization of foams from soy flour and concentrates. Studies indicated that removal of neutral lipids by hexane and polar lipids by aqueous alcohol markedly enhanced foaming properties (27, 28). Heating of soy isolates to 75-80 °C also enhanced soy protein foaming properties.

Soy Proteins as Additive Carriers

Certain metal ions, such as Ca, Mg, Fe and Na can bind to soy proteins by neutralizing charge effects. Binding of metal ions is generally driven by stabilization of protein structure, probably by allowing more rigid configuration of the molecule. Protein-mineral binding is a useful method for stabilizing proteins as well as minerals in food systems. Further, proteins can be used as a mineral carrier in food fortification. The presence of phytic acid (about 2.3 %) in soy protein inhibit the bioavailability of iron by forming stable iron-phytate complex (29). The iron bioavailability can be increased by binding iron to low molecular weight peptides produced from low phytate soy protein isolate. Low phytate (< 0.06 %) soy protein isolate has been produced by treating soy flour with the enzyme phytase followed by alkali solubilization (at pH 9.0) and isoelectric precipitation (at pH 4.5) (30). The iron content of this low phytate soy isolate was about 120 ppm, which was similar to that of regular soy protein isolate. However, low phytate soy protein isolate and peptides obtained from low phytate soy protein isolate had an iron binding capacity of 4000 ppm. Hence, these iron-peptide complexes could be used as bioavailable iron carriers.

The retention of food flavors during food processing and storage is a major problem, due to volatility of these flavors. The ability of soy proteins to bind flavors has been investigated by several researchers (31, 32, 33). Flavor binding may involve van der Waals, electrostatic, hydrogen bonding, or hydrophobic interaction between proteins and volatile low molecular weight flavor compounds. Proteins are used as carriers of flavors to prevent or minimize the loss of flavor during processing and storage.

Soy Protein-Hydrocolloid Complexes as Functional Ingredients

Protein-polysaccharide interaction is mainly electrostatic in origin, and the strength of the interaction is dependent on pH and ionic strength. Protein-polysaccharide interaction can be used to control protein solubility, to enhance gelation and stabilize emulsions and foams (34). In emulsion systems, interfacial protein-polysaccharide complexes produce more effective steric stabilizing layers around the oil droplets during emulsification and lead to prolonged storage stability (35). Proteins covalently linked to polysaccharide produce a hybrid macromolecular conjugate which has better emulsion stabilizing properties than either the pure proteins or polysaccharide (36).

Although soy proteins have the ability to form emulsions and foams, these are generally unstable during severe treatments such as pasteurization. Protein-polysaccharide complexes provide a simple solution to this problem (37). Increased viscosity of soy protein based emulsions and foams due to addition of xanthan gum contributed significantly to their high stability (Figures 1 & 2). The emulsions and foams produced by a 1:1 mixture of xanthan gum and soy proteins at a 0.2% total solid concentration were stable over a wide range of pH (3.0 to 9.0), ionic strength (0.1 to 1.0), and heat treatment (1 h at 85°C).

Structural Modification for Improving Soy Protein Functionalities

Because of inherent structural limitations, soy proteins lack the required physicochemical properties to meet the needs of various food and industrial uses. Modification of structural and conformational properties of proteins, to obtain optimum size, surface charge, hydrophobic/hydrophilic ratio, and molecular flexibility is a useful approach for improving desirable functional properties (38). Chemical and enzymatic methods have been used to modify protein structural features.

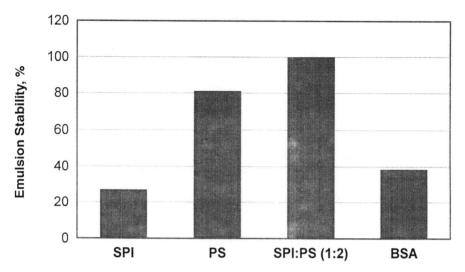

Figure 1. Emulsion Stability of Soy Protein Isolate (SPI), Xanthan Gum (XG), Bovine Serum Albumin (BSA), and SPI-XG Dispersion

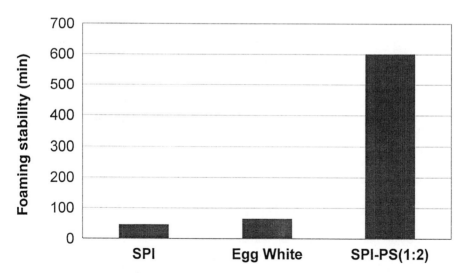

Figure 2. Foaming Stability of Soy Protein Isolate (SPI), Egg White, and SPI-Xanthan Gum (XG) Dispersion

Chemical Modification. Chemical modification such as alkali treatment, acylation, alkylation, phosphorylation, and deamidation (39, 40, 41) have been shown to improve functional properties of proteins. Acylation, phosphorylation and alkylation can modify amino acid side chains in proteins and lead to desirable structural and functional characteristics (39). Alkali treated soy protein has been used as paper coatings and binders in textile sizing (42). Alkali modification has improved adhesive and hydrophobic properties of soy protein (Table II) and hence would be very useful in producing wood glues from soy proteins with improved water resistance (43, 44, 45). However, alkali modification resulted in a high viscosity (>30,000 cP at 14 % solid concentration) adhesive. To alleviate this problem, a new method based on disulfide cleavage has been developed. Disulfide cleavage by 0.1 M Na_2SO_3 produced an adhesive with low viscosity (2,200 cP at 14% solid concentration), while having improved adhesive strength and hydrophobicity compared to unmodified soy proteins (Table II) (46). Further, data shown in Table 2 illustrate the differences in functionalities of spray dried and freeze dried products of modified soy proteins. The additional unfolding of modified soy proteins by heat treatment during the spray drying resulted in an increase in hydrophobicity and viscosity.

Table II. Adhesive Properties of Disulfide Cleaved Soy Proteins

Protein Samples	Adhesive Strength $(N/cm^2)^e$	Viscosity $(cP)^e$	Hydrophobicity ^{e,f}
SPI[a]	95[u]	240[v]	7[u]
AMSP[b]	180[v]	>30,000[x]	39[x]
MSPSD[c]	325[w]	2,200[w]	17[w]
MSPFD[d]	336[w]	100[u]	12[v]

a, unmodified soy protein isolate; b, Alkali modified soy proteins; c, spray dried product of modified soy proteins; d, freeze dried product of modified soy proteins; e, values in the same columns with different superscripts are significantly different from each other at $P<0.05$; f, measured using 1-anilino-8-naphthalene sulfonate method.

Soy protein based films prepared from a solution (6.25 % total solids at pH 9.0) containing soy protein isolate and wheat gluten in a ratio of 3:1, in the presence of 1% cysteine and 2% glycerol, had a three-fold increase in tensile strength when compared to soy protein films (Figure 3) (47). Intermolecular disulfide-sulfhydryl interchange reaction may be responsible for this increased tensile strength of soy protein-gluten film.

Enzymatic Modification. Enzymatic modification is an effective way to improve soy protein functional properties. Enzymes are used to cleave peptide bonds to produce peptides with desirable size, charge, and surface properties, or cross-link two protein molecules with complementary properties to improve protein functionalities. Enzymes, including pepsin, papain and trypsin, have frequently been used to enhance functional properties of soy proteins (12, 13). Modification of soy proteins by proteases improves foaming properties (3).

Enzymatic Hydrolysis. Peptic hydrolyzates of soy protein have enhanced foaming properties, and are used in confections, fudges, and meringues. Limited hydrolysis of soy proteins by papain resulted in a product having foaming properties similar to that of egg white (48). Smaller molecular size peptides produced by partial proteolysis have less tertiary, and possibly secondary, structure compared to the original proteins and exhibit increased solubility, decreased viscosity, and hence lead to significant changes in the foaming, gelling and emulsifying properties (Chobert, et al., 1988) which may contribute to desirable functional properties.

Soy protein hydrolysates obtained by incubating a 10 % soy protein isolate with papain (enzyme:protein ratio, 1:100) for 1 h at 37°C, had foaming capacity similar to that of egg white (22.0 vs. 21.2 ml) and significantly enhanced foaming stability compared to that of unmodified soy proteins (36.4 vs. 32.9) (Table III; Were et al., 1997). The foaming capacity of unmodified soy proteins was 18.0 ml. Papain modification also resulted in a two-fold increase in emulsifying activity of soy proteins (102 to 228 m^2/g) with significant increase in emulsion stability (36 to 47) (Table IV; Wu and Hettiarachchy, 1997). The solubility curve of papain modified soy proteins and unmodified soy proteins are compared in Figure 4. Papain modified soy protein exhibited higher solubility (100% at pH > 7.0) when compared to those of unmodified soy proteins (a gradual increase from 69% at pH 7.0 to 100 % at pH 9.0). Further, papain modification resulted in a five-fold increase in the hydrophobicity of soy proteins (40.8 vs 8.1). The enhanced solubility and hydrophobicity and the improved flexibility due to less secondary structure of smaller peptides were responsible for the improved functionality of papain modified soy proteins. While higher solubility promotes diffusion at the interface, enhanced hydrophobicity facilitates foam or emulsion formation. Improved flexibility of protein molecules helps to form stable interfaces, and hence enhance the foam or emulsion stability.

Table III. Foaming Properties of Papain Modified Soy Proteins

Sample	Foaming capacity (ml)[c]	Foam stability[c]
SPI[a]	18.0[u]	32.9[u]
PMSP[b]	22.0[w]	36.4[v]
Egg white	21.2[v]	45.9[w]

a, unmodified soy protein isolate; b, papain modified soy proteins; c, values in the same columns with different subscripts are significantly different from each other at P<0.05.

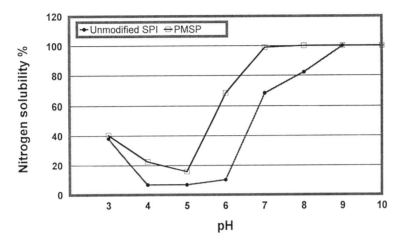

Figure 3. Solubility Curve of Soy Protein Isolate (SPI), and Papain Modified Soy Proteins (PMSP)

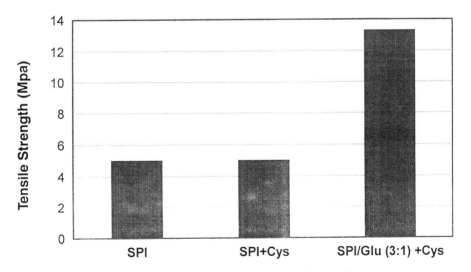

Figure 4. Tensile Strength of Soy Protein-Based Films

Table IV. Emulsifying Properties of Papain Modified Soy Proteins

Samples	Emulsion activity(m2/g)[c]	Emulsion stability[c]
SPI[a]	102[u]	36[u]
PMSP[b]	228[v]	47[v]

a, unmodified soy protein isolate; b, papain modified soy proteins; c, values in the same columns with different subscripts are significantly different from each other at P<0.05.

Although sodium dodecyl sulfate, which is used in cosmetic emulsions has high emulsion capacity, stability of these emulsions are very poor compared to protein emulsions. The stability of cosmetic emulsion formulation containing 0.2% sodium dodecyl sulfate (SDS) increased from 1.25 to over 20% when SDS was replaced up to 50% by papain modified soy proteins, while foaming capacity of cosmetic emulsion remained unchanged (Figures 5 and 6). Further increase in modified soy protein concentration caused a decrease in foam capacity.

Transglutaminase Cross-Linking. Cross-linking of casein with soy proteins (51, 52) and of whey proteins with soybean 11S proteins (53) using enzyme transglutaminase have been used to improve protein functionalities. Transglutaminase catalyzes acyl-transfer reactions introducing cross-linkages between proteins. New biopolymers were produced from a 1:1 mixture of whey proteins and 11S globulin by incubating with transglutaminase for 4 h at 37°C. The polymer product resulting from transglutaminase catalyzed cross-linking reaction was a mixture of heterologous and homologous biopolymers which showed enhanced foaming capacity and stability (Figure 7). Films made from this biopolymer had increased mechanical strength compared to soy protein films (Table V) (54). Increased interfacial interaction due to larger molecular size of cross-linked polymers could be responsible for this increased strength compared to soy protein films.

Plastein Reaction. Although proteolytic hydrolysis can improve solubility, emulsifying and foaming properties of soybean protein, bitter taste of peptides produced by hydrolysis limits their food uses. Plastein, a protein-like substance, can be produced from polypeptides by incubating the protein hydrolyzates with proteolytic enzymes under specific pH, temperature and substrate concentration. This reaction has been used to remove the undesirable bitter-taste and off-flavor, and incorporate deficient essential amino acids in protein products (31, 55). The plastein reaction is economically feasible and could be used for food, pharmaceutical and cosmetic applications (56). The mechanism of plastein reaction is still a controversial subject, which remains unresolved despite numerous efforts by several researchers. Yamashita el al. (57) reported that condensation or transpeptidation was involved in the production of plastein. However, several other researchers reported that plastein formation was merely a result of enzyme induced gelation from non-covalent interactions (58, 59). Data obtained in our

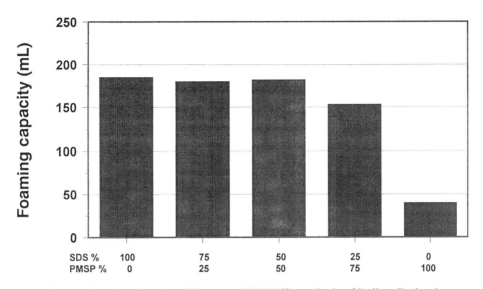

Figure 5. Foaming Capacity of Shampoo With Different Ratio of Sodium Dodecyl Sulfate (SDS) and Soy Protein Hydrolysate (SPH)

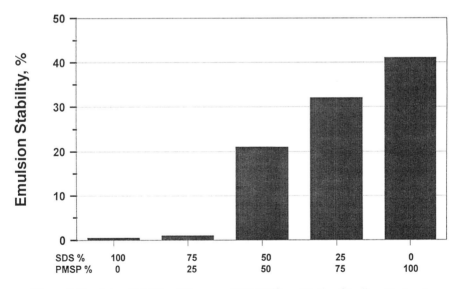

Figure 6. Emulsion Stability of Shampoo With Different Ratio of Sodium Dodecyl Sulfate (SDS) and Soy Protein Hydrolysate (SPH)

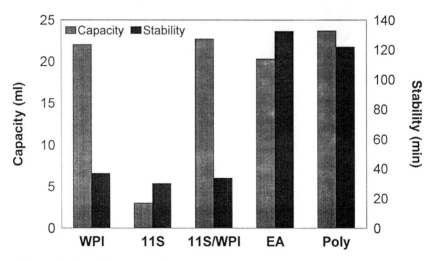

Figure 7. Foaming Properties Whey Protein Isolate (WPI), 11S, WPI/11S mixture, Egg Albumin, and Polymer

laboratory indicated no changes in molecular weight due to plastein reaction, indicating the involvment of non-covalent forces (60).

Table V. Properties of Films Made From Whey Protein, Soy 11S Protein and Whey Protein 11S Mixture and Those Treated With Transglutaminase

Film	Thickness (µm)	Water vapor permeability (g.mm/kPa.h.m2)	Tensile strength (MPa)	Puncture strength (N)
CWF	74a	2.71a	5.64a	5.81a
TWF	77a	4.88b	12.53b	9.07b
C11SF	69a	4.79b	7.61a	6.78a
T11SF	72a	7.84c	16.41c	9.29b
CMIXF	71a	4.91b	6.26a	6.09a
TMIXF	74a	7.68c	17.86c	10.01b

CWF: control whey protein film, TWF: film from whey protein treated with transglutaminase, C11SF: control 11S film, film from 11S treated with transglutaminase, CMIXF: control film from 1:1 mixture of 11S and whey protein, TMIXF: film from 1:1 mixture of 11S and whey protein treated with transglutaminase. Values in the same columns with different subscripts are significantly different from each other at P<0.05.

Future Research

Genetic engineering is becoming a more versatile tool for modifying plant seed storage proteins. Based on recombinant protein technology and knowledge of structural characteristics of soy proteins, it is possible to design soy protein genes to produce new proteins with desirable physicochemical, structural, and nutritional properties that are required for food and industrial applications. Biotechnology offers prospects of developing soybeans with better traits desired by users of soy protein products. Introduction of foreign genes into soybeans has been accomplished and is becoming a routine in some laboratories around the country. The design of proteins with appropriate structural and functional properties is a goal that may be achievable via recombinant DNA techniques. However, physical, chemical, and enzymatic modification of soy proteins still provide an important interim solution. There is a need for systematic research to acquire more fundamental information that can lead to soy protein products with desirable physicochemical properties for industrial and expanded food applications. Research in this direction should include: elucidation of molecular basis of functional properties; modification of protein structure; development of methodologies for studying structure-function relationships of soy protein in food and nonfood systems, development of novel functional soy protein-based ingredients.

Research and funding priorities should encourage soy protein researchers by assuring funding and the important of fundamental and applied research for producing functional soy protein ingredients.

Literature Cited

1. Damodaran, S. *In: Protein Functionality in Food Systems*; Hettiarachchy, N. S. and Ziegler, G. R. eds. Marcel Dekker, Inc. New York, NY, 1994, pp 1-37.
2. Morr, C. V. *J. Am. Oil Chem. Soc.* 1990, *67*:265-271.
3. Kinsella, J. E. *J. Am. Chem. Soc.* 1979, *56*:242-250.
4. Sorgentini, D. A.; Wagner, J. R.; Anon, M. C. *J. Agric. Food Chem.* 1995, *43*:2471-2479.
5. Lopez de Ogara, M.C.; Delgado de Layno, M.; Pilosof, A. M.; Macchi, R. A. *J. Am. Oil Chem. Soc.* 1992, *69*:184-187.
6. Nir, I.; Feldman, Y.; Aserin, A.; Garti, N. J. Food Science. 1994, 59:606-610.
7. Das, K. P.; Kinsella, J. E. Adv. Food Nutr. Res. 1990, 34:81-201.
8. Petruccelli, S.; Anon, M. C. *J. Agric. Food Chem.* 1996, 44:3005-3009.
9. Kinsella, J.E.; Fox, P. F. *CRC Crit. Rev. Food Science and nutrition.* 1986, 24:91.
10. Lambuth, A.L. In *Handbook of Adhesion*, Second edition, Skiest, I. S. ed. Ban Nostrand Reinhold, New York, 1977, pp 172-180 pp.
11. Shemer, M.; Creinin, H. L.; McDonald, R. E.; Irwin, W. E. *Cereal Chem.* 1978, 55:383.
12. Kim, S.Y.; Peter, S. W.; Rhee, K. C. *J. Agric. Food Chem.* 1990, 38:651-656.
13. Bernard-Don, L.S.; Pilosof, A. M. R.; Bartholomai, G. B. *J. Am. Oil Chem. Soc.* 1991, *68*; 102-105.
14. Phillips, L. G.; Whitehead, D. M.; Kinsella, J. E. *In Structure-Function Properties of Food Proteins*. Academic Press, Inc. Harcourt Brace and Company, Sandiego, 1994, pp 111-137, 179-204.
15. Utsumi S.; Kinsella, J. E. *J. Food Sci.* 1985, 50:1278-1282.
16. Grinberg, V.Y.; Grinberg, N. V.; Bikbov, T. M.; Bronich, T. K.; Mashkevish, A. Y. *Food Hydrocolloids*, 1992, 6:69-96.
17. Hermansson, A.M. *J. Text. Studies*, 1975, 5:425-439.
18. Kester, J.J.; Fennema, O. *Food Technol.* 1986, 40:47-59.
19. Krochta, J.M. In *Advances in Food Engineering*, Singh, R. P.; Wirakartakusumah, M. A. eds., CRV Press, London. 1992, pp 517-536.
20. Gennadios, A.; Mchugh, T. H.; Weller, C. L.; Krochta, J. M. In *Edible Coatings and Films to Improve Food Quality*. Krochta, J. M.; Baldwin, E. A.; Nisperos-Carriedo, M. O. eds. Technomic, Lancaster, PA, 1994, pp 201-277.
21. Nakai, S. *J. Agric. Food Chem.* 1983, 31:672-678.
22. Kinsella, J. E. *CRC Critical review in Food Sci. Nutr.* 1976, 7:219-249.
23. Hutton, C. W.; Campbell, A. M. 1977, *J. Food Sci.* 42:454-456.
24. Kinsella, J.E.; Whitehead, D. M. *ACS Symp. Ser. 343*, Brash, J. L.; Horbett. T. A. eds.Amer. Chem. Soc. Washington D.C. 1987.
25. Fleming, S. E.; Sosulski, F. W; Kilara, A.; Humbert, E. S. *J. Food Sci.* 1974, 39:188-191.
26. Lin, , M. J. Y.; Humbert, E. S.; Sosulski, F. W. 1974, *J. Food Sci.* 39:368-370.

27. Yatsumatsu, K.; Toda, J.; Wada, T.; Misaki, M.; Ishii, K. *Agric. Biol. Chem.* 1972, 36:537-543.
28. Eldridge, A. C.; Hall, P. K.; Wolf, W. J. *Food Technol.* 1963, 17:1592-1595.
29. Nelson, K. J.; Potter, N. N. J. Food Science. 1979, 44:106-109.
30. Kalapathy, U., Hettiarachchy, N. S. Unpublished data, 1997.
31. Arai, S.; Yamashita, M.; Fujimaki, M. *Cereal Foods World,* 1975, *20,* 107-112.
32. Franzen, K. L.; Kinsella, J. E. J. Agric. Food Chem. 1974, 22:675-678.
33. Damodaran, S.; Kinsella, J. E. J. Agric. Food Chem. 1981, 29:1249-1253.
34. Samant, S.K.; Singhal, R. S.; Kulkarni, P. R.; Rege, D. V. *Int. J. of Food Sci. and Tech.* 1993, 28: 547-562.
35. Dickinson, E. *J. Food Eng.* 1994, 22:59-74.
36. Dickinson, E; Galazka, V. B. In *Gums and Stabilizers for the Food Industry,* Vol. 6, Phillips, G. O.; Wedlock D. J.; Williams, P. A. eds. Oxford University Press, Oxford, UK. 1991, pp 351-362.
37. Xie, R; Hettiarachchy, N. S. J. Food Science, 1997, In press.
38. Feeney, R. E.; Whitaker, J. R. *In New proteins Foods,* Altschul, A. M.; Wilcke, J. L. eds. Academic Press, Inc., New York, 1985, pp 181-219.
39. Rhee, K.C. In *Proceedings of the World Congress on Plant Protein Utilization in Human Foods and Animal feedstuffs,* Applewhit, T. H. ed. AOCS, Champaign, IL, 1989, p. 23.
40. Sung, H.Y.; Chen, H. J.; Liu, T. Y.; Su, J. C. *J. Food Sci.* 1983, 48:716.
41. Hamada, J.S.; Marshall, W. E. *J. Food Sci.* 1989, 54:598-601.
42. Rhee K.C. *Inform,* 3:1044-1064 (1992).
43. Kalapathy, U.; Hettiarachchy, N. S.; Myers, D.; Hanna, M. A. *J. Am. Oil Chem. Soc.* 1995, 72:507-510.
44. Kalapathy, U.; Hettiarachchy, N. S.; Myers, D.; Rhee, K. C. *J. Am. Oil Chem. Soc.* 1996, 73:1063-1066.
45. Hettiarachchy, N.S.; Kalapathy, U.; Myers, D. *J. Am. Oil Chem. Soc.* 1995, 72:1461-1464.
46. Kalapathy, U.; Hettiarachchy; Rhee, K. C. *J. Am. Oil Chem. Soc.* 1997, 74:195-199.
47. Were, L.; Hettiarachchy, N. S.; Kalapathy, U. unpublished data, 1997
48. Were, L.; Hettiarachchy, N. S.; Kalapathy, U. *J. Food Science.* 1997, In press.
49. Chobert, J.M.; Bertrand-Harb, C.; Nicolas, M. G. *J. Agric. Food Chem.* 1988, 36:833-892.
50. Wu, W.; Hettiarachchy, N. S. unpublished data, 1997.
51. Motoki, M.; Nio, N.; Takinami, K. *Agric. Biol. Chem.* 1984, 48: 1257-1261.
52. Nio, N.; Motoki, M.; Takinami, K. *Agric. Biol. Chem.* 1985, 49: 2283-2286.
53. Yildirim, M.; Hettiarachchy, N. S.; Kalapathy, U. *J. Food Sci.* 1996, 61:1129-1131.
54. Yildirim, M.; Hettiarachchy, N. S. J. Food Science. 1997, submitted.
55. Yamashita, M.; Arai, S.; Tsai, S. J.; Fujimaki, M. *J. Agric. Food Chem.* 1971, 19:1151-1154.
56. Motoki, M.; Seguro, K. *Inform.* 1994, 5:308-313.
57. Yamashita, M.; Arai, S.; Tanimoto, S.; Fujimaki, M. *Agric. Biol. Chem.* 1973, 37:353-356.
58. Hartnett, E. K.; Satterlee, L. D. J. Food Biochem. 1990, 14:1-13.
59. Eriksen, S.; Fagerson, I. S. J. Food Sci. 1976, 41:490-493.
60. Qi, M.; Hettiarachchy, N. S. Unpublished data, 1997.

Chapter 7

Effect of Acylation on Flax Protein Functionality

F. Shahidi and P. K. J. P. D. Wanasundara

Department of Biochemistry, Memorial University of Newfoundland, St. John's, Newfoundland A1B 3X9, Canada

Flaxseed meal contains 34 to 37% protein on a dry weight basis. Proteins of the meal could be isolated by complexation with sodium hexametaphosphate; the yield of the protein isolated could be improved upon prior removal of mucilage from flaxseed. The protein isolate so prepared was acylated with acetic or succinic anhydride in order to improve its functional properties. The degree of acylation of free amino groups was progressively increased with increasing anhydride concentration, however, a lower degree of acylation was achieved when succinic anhydride was used in place of acetic anhydride. The color of the acylated proteins became lighter as the degree of acylation was increased. Both acetylation and succinylation increased solubility of the isolate at all levels of acylation examined. Emulsification properties of protein preparations were improved due to acylation, particularly for succinylated products. The foaming properties of flax protein isolates were not improved by acylation. Low degrees of acetylation improved fat binding capacity of flax protein isolates, but succinylation did not render such an effect. Acylation also increased surface hydrophobicity of the products and the highest value was observed at the lowest degree of acetylation.

Flaxseed has traditionally been cultivated for its linen and oil, but has gained renewed interest due to the beneficial health effects of its chemical constituents. The meal, which is a co-product of oil extraction, may serve as a potential source of plant protein for use in food products. Flaxseed contains 34 to 37% protein on a defatted, dry weight basis (*37*). Vassel and Nesbitt (*52*) were able to isolate the major flax protein and named it "linin". Madhusudhan and Singh (*29*) reported that globulins constituted 70-85% of flaxseed protein, two third of which had a molecular mass of 250 kDa and the remainder was low molecular weight in nature. The composition of amino acids of flax protein generally matches those of canola and soybean protein except that it is marginal in tryptophan. It is also high in its content of arginine, aspartic acid, glutamic acid and leucine as compared to canola and soy proteins (49, 54). Thus, flaxseed meal may be considered as a potential source of high quality plant protein for incorporation into food products. Flaxseed is the third largest oilseed crop produced in Canada after canola and soybean, however, flaxseed has not been a commercial source of plant protein, perhaps due to lack of scientific and technical information on its quality.

Functional Properties of Oilseed Proteins

Characteristics of oilseed proteins, that determine their food utilization, are collectively known as functional properties. Oilseed proteins provide economical ingredients for many formulated foods in the form of foam, emulsion or gel. Proteins are the preferred surface active and network forming agents in such foods. Functional properties denote the physico-chemical characteristics of proteins that determine their behavior in foods during consumption, processing, preparation, and storage. These physico-chemical properties and the manner in which proteins interact with other food components affect processing applications, quality and ultimately acceptance of food, both directly and indirectly. The type of functional property required for a protein or a protein mix varies with the particular food system in which it is present (27). Tables I and II list typical functional properties of oilseed proteins and their importance in food applications.

According to Kinsella (25, 26), functional properties of proteins are fundamentally related to their physical, chemical and structural/conformational characteristics. These include, size, shape, amino acid composition and sequence, charge and charge distribution, hydrophilicity/hydrophobicity ratio, secondary structures and their distribution (eg. α-helix, β-sheet and aperiodic structure), tertiary and quaternary arrangement of polypeptide segments, inter- and intra-subunit cross links (e.g. disulfide bonds) and the rigidity/flexibility of the protein in response to external conditions. According to Damodaran (9) functional properties of food proteins are manifestations of their hydrodynamic and surface related molecular properties. Viscosity and gelation are among hydrodynamic properties that are affected largely by the shape and size of the macromolecule. However, hydrodynamic properties are independent of amino acid composition and distribution of the proteins. Emulsifying and foaming properties, fat and flavor binding and solubility are surface-related properties that are largely affected by amino acid composition/distribution and molecular flexibility rather than size and shape of the macromolecule. Factors such as processing conditions, the method of isolation, environmental factors (temperature, pH and ionic strength, etc.) and interactions with other food components (carbohydrates, flavors, ions, lipids, proteins and water, etc.) alter the functional properties of a protein (26).

The fundamental relationships between conformational properties of food proteins and their functional behavior in food systems are poorly understood. There has been a continuous interest among food scientists to investigate the molecular basis for the expression of functional properties of food proteins which helps to increase utilization of novel food proteins such as those from oilseed meals in conventional foods. Therefore, it is necessary to develop better processing techniques to retain or enhance protein functionality. Also different strategics such as chemical and enzymic

Table I. Functional properties of oilseed proteins important in food applications.[1]

General property	Specific functional attribute
Hydration	Wettability, water absorption, water-holding capacity, swelling, solubility, thickening, gelling, syneresis
Kinesthetic	Texture, mouthfeel, smoothness, grittiness, turbidity, chewiness
Sensory	Color, flavor, odor
Structural and rheological	Viscosity, elasticity, adhesiveness, cohesiveness, stickiness, dough formation, aggregation, gelation, network formation, extrudability, texturizability, fibre formation
Surface	Emulsification, foaming/aeration/whipping, protein-lipid film formation, lipid binding, flavour binding
Other	Compatibility with other food components, enzymatic activity, antioxidant properties

[1]Adapted from reference 27.

Table II. Typical functional properties conferred to foods by oilseed proteins.[1]

Functional property	Mode of action	Food system
Cohesion-adhesion	Protein acts as an adhesive material	Meats, sausages, baked goods, cheese, pasta products
Elasticity	Hydrophobic bonding in gluten, disulphide links in gels	Meats, bakery products
Emulsification	Formation and stabilization of fat emulsions	Sausages, bologna, soup, cakes
Fat absorption	Binding of free fat	Meats, sausages, doughnuts
Flavour binding	Adsorption, entrapment, release	Simulated meats, bakery goods
Foaming	Formation of stable films to entrap gas	Whipped toppings, chiffon desserts, angel food cakes
Gelation	Protein matrix formation and setting	Meats, curds, cheese
Solubility	Protein solvation	Beverages
Viscosity	Thickening, water binding	Soups, gravies
Water absorption and binding	Hydrogen bonding of water, entrapment of water without dripping	Meats, sausages, bread, cakes

[1]Adapted from references *26* and *27*.

modification or genetic engineering are needed to alter the conformational characteristics of underutilized food proteins in order to improve their functionality.

Modification of Functional Properties of Oilseed Proteins

Oilseed proteins in the native state do not always meet the desirable functional properties for particular food systems. Therefore, altering the functional properties of proteins by using biological, chemical or physical modification may be necessary. The food industry considers desired functional properties of new sources of proteins to include solubility characteristics, water and fat binding properties, foaming, emulsifying and viscoelastic properties. Modification of proteins involves changes in protein structure or conformation at all the primary, secondary, tertiary and quaternary levels. Besides, modification of functional properties, retarding deteriorative reactions (eg. Maillard reaction), removing of toxic or inhibitory ingredients (eg. phytic acid, phenolic acids and tannins) and attachment of nutrients and additives by formation of new covalent bonds (eg. amino acids, carbohydrates, flavor compounds and lipids) can also be achieved by biological or chemical modification of proteins (*13, 34*).

Chemical Modification

Chemical modification of proteins has been widely used in studies of structure-activity relationships. This technique has also been applied to study the structure of food proteins as well as to improve their nutritional and functional properties. Some of the most commonly used methods of chemical modification of amino acid residues are listed in Table III. Treatments which modify one or more amino acids and involve pH modification or chemical changes may confer desired functional properties to protein products. Most studies on food protein modification involve derivatization of ϵ-amino group of lysine residues. This type of derivatization may directly affect the net charge and charge-density of protein molecules. Conformational changes and alterations in intra- and intermolecular interactions may also occur that modify effective hydrophobicity of proteins (*14,31,34*). Since reactive side chains of proteins are nucleophilic (e.g. amino, thiol and phenolic hydroxyl) or may be easily oxidized or reduced, reactions involving modification of these groups are commonly carried out.

Acylation

The formation of an amide (isopeptide) bond, particularly with lysine residues, using an acid anhydride is the commonly used method of chemical modification for food/plant proteins (*13*). Reaction with acetic anhydride (acetylation) replaces the positively charged ϵ-amino group of the lysine residue by a neutral acetyl group (Reaction 1). Acylation with cyclic anhydrides such as succinic anhydride results in a two-charge change, from the positive charge of the ϵ-amino group to the negative charge of succinyl anionic groups (Reaction 2). The double-charge change of amino group to the carboxyl group due to succinylation increases the net negative charge and has a marked effect on protein conformation (*22, 31*).

Table III. Chemical modification of amino acid side chains.[1]

Side chain	Amino acid residue	Commonly used modification
Amino	Lysine	Alkylation, acylation
Carboxyl	Aspartic and glutamic acid	Esterification, amide formation
Disulphide	Cystine	Reduction, oxidation
Sulphydryl	Cysteine	Alkylation, oxidation
Thioether	Methionine	Alkylation, oxidation
Imidazole	Histidine	Alkylation, oxidation
Indole	Tryptophan	Alkylation, oxidation
Phenolic	Tyrosine	Acylation, electrophilic substitution
Guanidino	Arginine	Condensation with dicarbonyls

[1]Adapted from reference 34.

Reaction 1

$$\text{Protein-NH}_2 \; + \quad \begin{matrix} \text{O} \\ \parallel \\ \text{CH}_3\text{C} \\ \diagdown \\ \quad\quad \text{O} \\ \diagup \\ \text{CH}_3\text{C} \\ \parallel \\ \text{O} \end{matrix} \quad \longrightarrow \quad \begin{matrix} \text{O} \\ \parallel \\ \end{matrix} \text{Protein-NH-C-CH}_3 \; + \; \text{CH}_3\text{COO}^- \; + \; \text{H}^+$$

Reaction 2

$$\text{Protein-NH}_2 \; + \quad \begin{matrix} \text{O} \\ \parallel \\ \text{CH}_2\text{C} \\ | \quad\quad \diagdown \\ \quad\quad\quad \text{O} \\ | \quad\quad \diagup \\ \text{CH}_2\text{C} \\ \parallel \\ \text{O} \end{matrix} \quad \longrightarrow \quad \begin{matrix} \text{O} \quad\quad \text{O} \\ \parallel \quad\quad \parallel \\ \end{matrix} \text{Protein-NH-C-CH}_2\text{CH}_2\text{-C-O}^- \; + \; \text{H}^+$$

Acylation reactions are known to follow the carbonyl addition pathway. While amino and tyrosyl groups are easy to acylate, histidine and cystine residues are only rarely observed to undergo acylation. The hydroxyl groups of serine and threonine are weak nucleophiles and never acylate in the aqueous medium. Acylation reactions are generally influenced by the pH of the medium. Table IV summarizes the effect of acylation on functional properties of some oilseed proteins. Acetylation and succinylation of plant proteins have been reported to increase protein solubility (28, 38-41), improve emulsifying and foaming properties (7), increase water holding and oil binding capacities (46) and improve flavor quality (15-16) of plant protein products. According to Thompson and Cho (50) and Ponnampalam et al. (46) succinylation enhances nitrogen extractability and emulsifying properties of canola flour. Protein isolates prepared from acylated canola flour contained low phytic acid but the yield of protein recovery was low (51).

Flaxseed meal exhibits a very good water absorption, emulsifying and foaming properties due to the chemical nature of its proteins and polysaccharides. Dev and Quensel (11) have reported that functional properties of alkali-extracted flax proteins were comparable to those of soy protein isolates. Heat treatment improved water absorption but reduced fat absorption, nitrogen solubility and foaming as well as emulsifying properties of the isolated flaxseed proteins (30). However, there are no reports on chemical modification of flaxseed proteins or on the functionality of such preparations.

Flaxseed proteins could be isolated by complexation with sodium hexametpho-sphate (55). The yield of protein isolation could be increased by using mucilage-reduced flaxseed meal (56). The flax protein isolate used for this study was prepared from low mucilage flaxseed obtained by soaking in 0.10 M Na_2CO_3 solution for 12 h via sodium hexametaphosphate (SHMP) complexation (55) and contained 79% protein.

Amino Acid Composition

The amino acid composition of the prepared protein isolate is presented in Table V. The contents of individual amino acids in flax protein isolates were within the range reported in previous studies (6, 54). Tryptophan and lysine were the first limiting amino acids of flaxseed and its protein isolate, respectively. The content of lysine in the isolate was lower than that in the seed, similar to that reported by Sosulski and Sarwar (49) for flax protein isolates prepared by alkali solubilization. The percentage ratios of essential to total amino acids (%E/T) for the seed and protein isolate were well above 36% reported for ideal proteins as recommended by FAO/WHO (12).

Acylation of Flaxseed Protein Isolate

Treatment of flaxseed protein isolate with increasing quantities of acetic or succinic anhydride progressively acylated the free amino groups of seed proteins. The degree of acylation of flax protein isolate was determined as percentage loss of ϵ-amino groups of lysine residues (Table VI). Due to the high nucleophilic character and steric hindrance the ϵ-amino group of lysine residues are most reactive towards acylation, thus the degree of protein modification is commonly expressed as the percentage of blocked amino groups of lysine. Compared to succinylation, a higher degree of acetylation was observed at the three concentrations of anhydride used (Table VI). Nitecka et al. (36) have also reported that acetic anhydride is more effective than succinic anhydride in blocking ϵ-amino groups of rapeseed proteins. Results of this study indicated that all free amino groups of isolated flaxseed proteins were not acylated. Incomplete acylation of reactive amino acid residues by an acetyl, succinyl or any other acyl group is common for storage proteins of oilseeds (7, 15, 46, 47).

Effect of Acylation on Functional Properties

Structural changes of proteins such as those in molecular mass (dissociation), shape (unfolding) and charge due to acetylation or succinylation have been used to explain variations in functional properties of chemically modified storage proteins of seeds. These alterations change hydration properties and surface activities of proteins and affect their solubility and surface activity-related properties (emulsifying and foaming) of the native proteins.

Color

The flaxseed protein isolate prior to acylation had a light yellow color and a fluffy texture. The isolate became more whitish due to acylation. The Hunter color values, measured as described by Wanasundara and Shahidi (57), indicated that the Hunter L value increased (68.8 to 74.2) with the increased degree of acylation which indicated a lighter color for the products compared to the unmodified isolate. The increase in Hunter L value was higher at the highest degree of succinylation (56.9%) than the highest degree of acetylation (84.5%) (Table VII). A decrease in Hunter b values of the acylated products indicates a decrease in their yellow color which was also

Table IV. Acylation of oilseed proteins and the effect on functional properties.

Protein source[1]	Modification reaction	Changes in chemical and physical properties	Reference
Cottonseed	Acetylation	Improved water holding, oil binding, emulsifying and foaming capacities.	7
	Succinylation	Improved water solubility, heat stability, emulsion and oil binding capacities, gel strength, viscosity, water hydration and retention, decreased bulk density.	7, 8
Canola	Succinylation	Improved solubility, emulsifying activity and stability, thermally induced gelation, hydrophobicity and net negative charge.	38, 39, 40, 41
Rapeseed	Acetylation	Improved nitrogen solubility, emulsifying properties, specific viscosity.	46
	Succinylation	Improved emulsifying properties, foam capacity and stability, heat stable gel formation, dissociation into subunits and unfolding.	18, 47

Rapeseed flour	Acetylation	Decreased phenolic acid extractability and peptic and tryptic hydrolysis, increased nitrogen extractability, decreased phytic acid and mineral extractability.	*45, 49*
	Succinylation	Increased nitrogen extractability, decreased phytic acid and mineral extractability.	*50, 51*
Peanut	Succinylation	Enhanced solubility at low pH, water absorption and viscosity, dissociation into subunits and swelling.	*5*
Soybean	Acetylation	Decreased water binding and gel strength, increased solubility.	*4, 15*
	Succinylation	Increased emulsifying activity and stability, foam capacity, decreased isoelectric pH.	*15*

[1]Isolates, unless otherwise specified.

Table V. Amino acid composition (g/16g N) of low-mucilage flaxseed and flax protein isolate prepared via sodium hexametaphosphate complexation.[1]

Amino acid	Low-mucilage flaxseed (0.10 M NaHCO$_3$, 12 h soaked)		Standard protein[2]
	Meal	Protein isolate	
Essential amino acids			
Histidine	2.50 ± 0.11	2.18 ± 0.02	-
Isoleucine	4.54 ± 0.30	4.41 ± 0.48	4.00
Leucine	6.54 ± 0.26	6.28 ± 0.53	7.04
Lysine	4.55 ± 0.18	3.62 ± 0.06	5.44
Methionine+Cysteine	6.09 ± 0.02	5.06 ± 0.07	3.52
Phenylalanine+Tyrosine	7.50 ± 0.08	6.90 ± 0.01	6.08
Threonine	4.37 ± 0.20	4.79 ± 0.08	4.00
Tryptophan	0.70 ± 0.02	1.55 ± 0.03	0.96
Valine	5.46 ± 0.06	4.29 ± 0.05	4.96
Non-essential amino acids			
Alanine	5.02 ± 0.10	4.54 ± 0.06	-
Aspartic acid+asparagine	10.65 ± 0.43	10.33 ± 1.20	-
Arginine	10.03 ± 0.50	10.21 ± 1.01	-
Glycine	7.44 ± 0.82	5.54 ± 0.08	-
Glutamic acid+glutamine	20.76 ± 1.09	19.99 ± 2.18	-
Proline	3.98 ± 0.23	3.77 ± 0.10	-
Serine	5.02 ± 0.09	4.67 ± 0.09	-
Total essential amino acids[3]	39.75	36.90	36.00
E/T %	37.80	37.60	-

[1] Results are mean values of three samples ± standard deviation.
[2] From reference *12*
[3] Histidine not included

Table VI. Degree of acylation of flaxseed protein isolate.[1]

Added anhydride (g/g protein equivalent)	Degree of acylation (%)
Acetic anhydride	
0.05	47.7 ± 2.0
0.10	70.0 ± 4.0
0.20	84.5 ± 2.5
Succinic anhydride	
0.01	38.2 ± 2.8
0.10	45.3 ± 3.5
0.20	56.9 ± 4.0

[1] Results are mean values of duplicate determinations of three samples ± standard deviation.
Adapted from reference *57*.

Table VII. Hunter L and b color values of modified and unmodified flax protein isolates.[1]

Treatment	L	b
Unmodified	68.8±0.8[a]	26.5±0.9[c]
Modified		
Acetic anhydride (g/g protein equivalents)		
0.05	69.2±1.0[a]	23.6±0.8[ab]
0.10	70.0±2.2[ab]	23.6±0.9[ab]
0.20	71.8±1.0[b]	22.3±0.9[ab]
Succinic anhydride (g/g protein equivalents)		
0.05	72.0±0.9[b]	23.9±0.8[ab]
0.10	72.9±1.2[bc]	22.6±1.1[a]
0.20	74.2±0.8[c]	21.3±0.8[a]

[1]The Colorimet unit was calibrated using a white tile with Hunter values of $L=94.5\pm0.2$, $a=-1.0\pm0.1$, $b=0.0\pm0.2$.
Results are mean values of duplicate determinations of three samples ± standard deviation. Means followed by different superscripts within a column are significantly ($p<0.05$) different from one another.
Adapted from reference 57.

Table VIII. Fat binding capacity of acylated flax protein isolates.[1]

Treatment	Fat binding capacity (ml/100g)
Unmodified	93.0 ± 1.7[d]
Modified	
Acetic anhydride (g/g protein equivalents)	
0.05	105.0 ± 2.1[f]
0.10	98.7 ± 1.6[e]
0.20	94.7 ± 1.3[d]
Succinic anhydride (g/g protein equivalents)	
0.05	82.5 ± 1.9[c]
0.10	72.6 ± 1.4[b]
0.20	63.9 ± 1.5[a]

[1]Mean ± SD of three samples.
Means followed by different superscripts within a column are significantly ($p<0.05$) different from one another.
Adapted from reference 57.

noticeable by the naked eye. Flax protein isolates with the highest degree of succinylation (56.9%) had the highest Hunter **L** and the lowest Hunter **b** values and were almost white in color. The Hunter **a** values of the protein products did not show any specific pattern of change (data not shown). No off-odors were detected in any of these products.

The proteins participate in the color properties of foods by light scattering effects resulting in product lightness (whiteness), participating in browning and enzymatic reactions and also serving as a carrier/ ligand for prosthetic chromaphores (*1*). According to Habeeb *et al.* (*19*) replacement of short range attractive forces (ammonium-carboxyl) with short range repulsive forces (succinate carboxyl-carboxyl) due to succinylation may alter the molecular conformation of protein molecules. The resultant negative charge, in combination with the electrostatic repulsion due to the introduction of succinate anion, causes looser texture, higher bulk density and lighter color of the succinylated proteins than their unmodified counterparts. Light colored products have been obtained from succinylation of fish (*17*), leaf (*16*), and soybean (*32*) proteins. The exact mechanism by which this change is brought about remains unknown. When prepared protein isolates are dehydrated and milled product whiteness increases due to increased light scattering by particles having size between 0.1 and 1.0 μm in diameter (*1*). According to Franzen and Kinsella (*15-16*), inclusion of water molecules is facilitated due to the repulsion of adjacent polypeptide molecules thus increasing the solubility of succinylated seed proteins.

Solubility

Figure 1 represents the percentage solubility changes of acetylated and succinylated protein isolates under different pH and NaCl concentrations of the medium. At low degree of acylation both acetylated (Figure 1B, C and D) and succinylated (Figure 1E, F and G) products had a similar solubility pattern. Succinylated products had the highest solubility values at all degrees of modifications studied. A decreased solubility was observed at acidic (low) pH values for all protein isolates and their modified products. However, succinylated (Figure 1E, F and G) products showed higher solubilities in the acidic pH range as compared to acetylated (Figure 1B, C and D) and unmodified (Figure 1A) protein isolates. Increased concentration of NaCl had a negative effect on the solubility of modified proteins especially acetylated products which was clearly observed at high pH values. Compared to the acetylated or unmodified proteins, succinylated products were soluble even in 0.70M NaCl solutions. Flax protein isolate showed similar patten of solubility with pH change as reported for some other acylated plant protein isolates (*15,23,38,48*).

Studies of Gueguen *et al.* (*18*) on rapeseed globulin suggest that succinylation induces a stepwise dissociation of the 12S globulin and leads to the unfolding of its subunits. The combination of intra- and intermolecular charge repulsion also promotes protein unfolding and induces fewer protein-protein and more protein-water interactions. The unfolding of the protein molecule, and dissociation into subunits, shifts the isoelectric point of the proteins to lower values, thus making the acylated proteins more soluble in the acidic pH range as compared with their unmodified counterparts. As the net negative charge is proportional to the extent of succinylation, solubility of flax

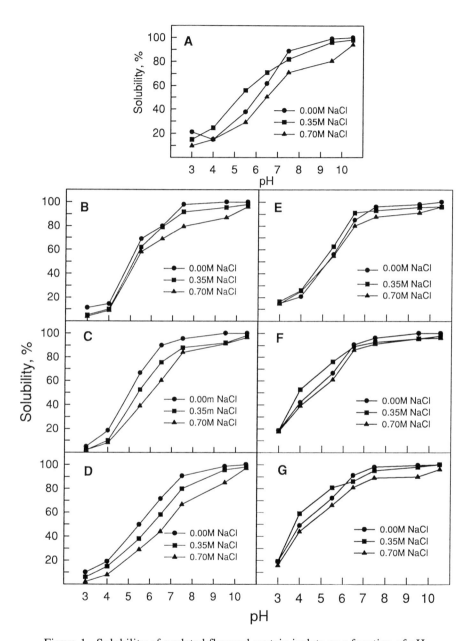

Figure 1. Solubility of acylated flaxseed protein isolate as a function of pH and NaCl concentration: (A) unmodified, (B) 0.05, (C) 0.10, (D) 0.20 g acetic anhydride/g protein equivalent and (E) 0.05, (F) 0.10, (G) 0.20 g succinic anhydride/g protein equivalent. (Adapted from reference 57).

protein isolates would increase as the number of succinylated groups increases (Figure 1E, F and G). Solubility is a very important functional property of proteins as it is a critical prerequisite for using proteins in beverages, liquid foods and for emulsion and foam formation. In fact, solubility reflects the balance of charge and hydrophobicity of the proteins (21) which affects their interaction with the solvent and other protein molecules (35).

Acetylation reduces the extent of electrostatic attraction between neighboring cationic amino and anionic carboxyl groups due to substitution of amino groups with neutral acetyl groups. As a result, acetylated flax proteins behave differently from succinylated products and exhibited lower solubility when compared to their succinylated counterparts.

Emulsifying Properties

The emulsifying activities (EA) of unmodified flax protein isolate, measured according to the method of Paulson and Tung (39) were increased as the pH and concentration of NaCl in the medium increased. The emulsifying activities of acetylated and succinylated flax protein isolates, as affected by the pH change and NaCl concentration of the medium, are presented in Figure 2. The difference in the EA for acetylated and succinylated proteins indicates that the effect of pH and NaCl concentration on EA was different for the acyl group involved in modification of protein. Among the acetylated products EA showed a decrease when the degree of acetylation increased from 0.0 to 84.5% (Figure 2B, C and D). The effect of NaCl concentration was marginal.

Succinylation showed an increasing influence on the EA of protein isolates (Figure 2E, F and G). As the degree of succinylation increased a higher EA was observed, in contrast to the acetylated flax protein isolates. Increase in pH resulted in an increase in EA, but an increase in the NaCl concentration in the medium gave lower EA values for succinylated flax protein products, especially at the lowest degree of succinylation studied (Figure 2E). It is clear that as the degree of succinylation increased there was an improvement in the EA at pH values between 5 and 8 (Figure 2E, F and G).

The emulsion stability (ES) of the prepared products is given in Figure 3 as half-life of the emulsion (time required to reduce the absorbance at 500 nm by 50%) for acetylated and succinylated derivatives of the flax protein isolate. Emulsion stability increased with increasing the degree of acetylation (Figure 3B and C) and succinylation (Figure 3E and G), except at the highest degrees of acylation (Figure 3D and 3G). As the pH increased ES was improved; however, increased NaCl concentration in the medium did not give a similar effect for the acylated products.

Succinylation has been reported to improve the emulsifying properties of oilseed proteins (7, 15-16, 39, 50-51). As a reflection of increased solubility and looser structure of succinylated proteins, diffusion/migration of the protein molecules to the oil-water interface and rearrangement within the interfacial film is facilitated (59). Therefore, good solubility of a protein is essential for formation of emulsions (34). Pearce and Kinsella (42) reported that succinylation of yeast proteins affords smaller droplets in oil-in-water emulsions as compared to unmodified proteins; the decrease in droplet size increases interfacial area and hence increases the emulsifying ability of

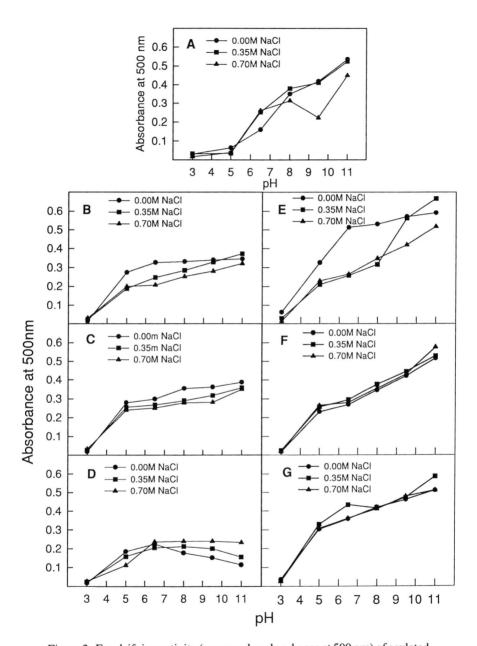

Figure 2. Emulsifying activity (measured as absorbance at 500 nm) of acylated flax protein isolates as a function of pH and NaCl concentration: (A) unmodified, (B) 0.05, (C) 0.10, (D) 0.20 g acetic anhydride/g protein equivalent and (E) 0.05, (F) 0.10, (G) 0.20 g succinic anhydride/ g protein equivalent. (Adapted from reference 57).

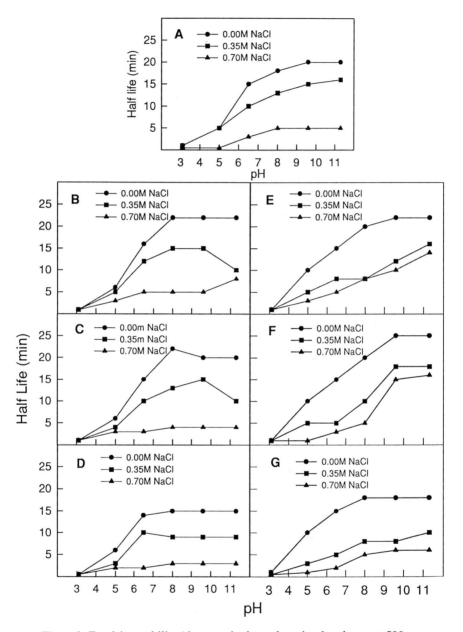

Figure 3. Emulsion stability (time required to reduce the absorbance at 500 nm by 50%) of acylated flax protein isolates as a function of pH and NaCl concentration: (A) unmodified, (B) 0.05, (C) 0.10 (D) 0.20 g acetic anhydride/g protein equivalent and (E) 0.05, (F) 0.10, (G) 0.20 succinic anhydride/g protein equivalent. (Adapted from reference 57).

proteins. Watanabe and Arai (*60*) have also reported that increased surface area increases surfactant properties of proteins. As the protein becomes more soluble, it forms layers around the fat globule and facilitates association with the aqueous phase which encloses the fat globule and renders the emulsion more stable and resistant to coalescence (*20*). A direct relationship exists between the solubility and ability of succinylated plant proteins to emulsify oil (*15-16,46,59-60*). However, emulsifying properties of acylated proteins do not depend solely on solubility. The hydrophilic-lipophilic balance (HLB) of a particular protein is also important, but does not necessarily increase linearly with the increase in protein hydrophobicity (*33*). Thus, acetylated proteins do not show better emulsifying properties although they exhibit higher solubility when they are compared with their unmodified counterparts. The unfolding of the protein structure due to succinylation may expose more hydrophobic groups buried in the molecule and could change its hydrophobicity and hence the HLB value. Halling (*20*) has suggested that increased rheological strength of protein films could reduce mechanical deformation and desorption of the interfacial proteins to give more emulsified droplets. Therefore, emulsion formation could be favored by protein-protein (hydrophobic) interaction as well as rheological properties of the interfacial protein film which encapsulates the oil droplets (*10*).

Fat Binding Capacity

The fat binding capacity (FBC) of flax protein products are presented in Table VIII. Unmodified flax protein isolate was able to bind 93.0 ± 1.7 ml of corn oil per 100 g of material. Acylation changed FBC of the flax protein isolate. The highest FBC was observed at the lowest degree of acetylation (47.7%) and the lowest value was obtained at the highest degree of succinylation (56.9%). Succinylated protein products (at all degrees of succinylation) had lower FBCs than the acetylated products.

Fat binding capacity of proteins is important as it enhances flavor retention and improves mouthfeel. The mechanism of fat absorption of a protein was suggested as physical entrapment of oil by protein particles (*58*). However, the method used in this study was proposed by Voutsinas and Nakai (*53*) in order to minimize any physical entrapment effect and to estimate only the absorbed fat by the protein. The protein-lipid interactions are affected by protein conformation, protein-protein interaction and the spatial arrangement of the lipid phase resulting from lipid-lipid interaction. Non-covalent interactions such as hydrophobic, electrostatic and hydrogen bonding forces are also involved in protein-lipid interactions. The aromatic hydrophobicity indicates the exposed aromatic (hydrophobic) residues of the protein (*10*).

Foaming Properties

The foaming capacity and foam stability of flax protein isolates are presented in Table IX. The highest foaming capacity was observed for the unmodified protein isolate and the isolates with the lowest degree of modification. As the degree of acetylation or succinylation increased the foaming capacity decreased and isolates with the highest degree of acylation had the lowest foaming capacity. The foams of acylated flax protein isolates were less stable than unmodified isolates (Table IX). The foam

Table IX. Foaming properties of modified and unmodified flax protein isolates.[1]

Treatment	Foam expansion $(\%)^2$	Foam stability $(\%)^3$
Unmodified	112.0±5.0[d]	51.0±3.0[e]
Modified		
Acetic anhydride (g/g protein equivalents)		
0.05	97.0±2.5[d]	36.1±2.0[c]
0.10	92.5±1.0[c]	28.5±1.0[b]
0.20	74.5±1.0[b]	14.6±1.0[a]
Succinic anhydride (g/g protein equivalents)		
0.05	98.2±2.0[d]	44.3±1.0[d]
0.10	88.1±3.4[c]	40.3±1.4[c]
0.20	66.9±2.0[a]	16.0±2.0[a]

[1]Mean ± SD of three samples. Means followed by different superscripts within a column are significantly ($p<0.05$) different from one another. Adapted from reference 57.

[2] Volume increase after whipping 50 ml of 1% (w/v) protein solution.

[3]Foam remained after 15 min as a percentage of original foam volume.

stabilities of the modified protein isolates were parallel to the pattern of foaming capacity.

A good foam-forming protein should be rapidly adsorbed at the air-water interface during whipping, reduce the surface and interfacial tension of the liquid and form a structural, continuous, cohesive film around air bubbles (*10, 25*). Increased negative charge of the protein could hinder protein-protein interactions due to changes in structure and hydrophobicity and may lead to low foaming ability of the modified protein products; however, solubility is positively related to foaming ability. Foam stability is affected by the factors including rheological properties such as viscosity and shear resistance, film elasticity and the magnitude of disjoining pressure between the protein layers. These factors affect the foam stability by affecting liquid drainage from the lamellar film. The rheological properties of protein films that contribute to foam stability relate to intermolecular interactions and mechanical strength (*10*). Gueguen *et al.* (*18*) have suggested that unfolding of the protein molecule due to succinylation may increase both the viscosity and interactions between polypeptide chains that may overcome the repulsive effects of the negative charges. Therefore, better foam forming and stabilizing effects of highly succinylated proteins could be related to electrostatic repulsive forces between the air bubbles which are due to the increasing charge of the adsorbed proteins. However, such behavior was not observed for the succinylated flax protein isolates.

Surface (Aromatic) Hydrophobicity

Surface hydrophobicity of flax protein isolates, measured as fluorescence intensity/mg protein, is presented in Table X. Acetylation brought about an increase in surface hydrophobicity. Thus, as the degree of acetylation increased the hydrophobicity values also increased.

Surface hydrophobicity of a protein is related to the extent of hydrophobic patches on the protein surface (*10*). Exposure of the hydrophobic interior and modification of positively charged lysine residues with uncharged acetyl groups results in a decrease in net charge of proteins and hence increased surface hydrophobicity (*24*). Unfolding of the protein molecules may also make hydrophobic sites accessible for binding the fluorescence probe 1-anilino-8-naphthalene sulfonate (ANS). Surface and molecular hydrophobicity of protein molecules have been shown to be related to the solubility and emulsifying properties of cowpea (*2-3*) and soy (*43-44*) globulins. Schwenke *et al.* (*47*) have reported an increased surface hydrophobicity of pea proteins at low and moderate levels of succinylation. According to these authors a drop in surface hydrophobicity was observed after passing the critical level (<70%) of N-succinylation, perhaps due to the effect of high negative charge density that inhibits the ANS molecule from approaching and binding to the protein surface.

Summary

The color and solubility of flax protein isolates were improved due to acetylation and succinylation. Succinylated protein products had better emulsifying activities than their acetylated or unmodified counterparts. Presence of NaCl in the medium had a marked

Table X. Surface hydrophobicity of acylated flax protein isolates.[1]

Treatment	Surface hydophobicity (fluorescence intensity/mg protein)
Unmodified	200 ± 25
Modified Acetic anhydride (g/g protein equivalents)	
0.05	450 ± 31
0.10	378 ± 22
0.20	352 ± 18
Succinic anhydride (g/g protein equivalents)	
0.05	402 ± 19
0.10	318 ± 18
0.20	249 ± 15

[1] Mean $\pm$ SD of three samples.
Adapted from reference 57.

effect on the solubility and emulsifying activity of acetylated products. Acylation did not improve emulsion stability or foaming properties of flax protein isolates. Low degrees of acetylation enhanced the fat binding capacity of the flax seed protein.

Literature Cited

1. Acton, J.C.; Dawson, P.L. Color as a Functional property of proteins. In *Protein Functionality in Food Systems*; Hettiarchchy, N.S. and Zeigler, G.R.; Eds.; Marcel Dekker, Inc. New York, NY. 1994. pp. 357-381.
2. Aluko, R.E; Yada, R.Y. Relationship of hydrophobicity and solubility with some functional properties of cowpea (*Vigna unguiculata*) protein isolate. *J. Sci. Food Agric.* **1993**, *62*, 331-335.
3. Aluko, R.E; Yada, R.Y. Structure-function relationships of cowpea (*Vigna unguiculata*) globulin isolate: influence of pH and NaCl on physicochemical and functional properties. *Food Chem.* **1995**, *53*, 259-265.
4. Barman, B.G.; Hansen, J.R.; Mossey, A.R. Modification of the physical properties of soy protein isolate by acetylation. *J. Agric. Food Chem.* **1977**, *25*, 638-641.
5. Beuchart, L.R. Functional and electrophoretic characteristics of succinylated peanut flour. *J. Agric. Food Chem.* **1977**, *25*, 258-261.
6. Bhatty, R.S. and Cherdkiatgumchai, P. Compositional analysis of laboratory-prepared and commercial samples of linseed meal and of hulls isolated from flax. *J. Am. Oil Chem Soc.* **1990**, *67*, 79-84.
7. Child, E. A.; Parks, K. K. Functional properties of acylated glandless cottonseed flour. *J. Food Sci.* **1976**, *41*, 713-714.
8. Choi, Y. R.; Lusas, E. W.; Rhee, K.C. Succinylation of cotton seed flour: Effect on the functional properties of protein isolates prepared from modified flour. *J. Food Sci.* **1981**, *46*, 954-955.
9. Damodaran, S. Interrelationship of molecular and functional properties of food proteins. In *Food Proteins*; Kinsella, J.E.; Soucie, W.G., Eds.; American Oil Chemists' Society, Champaign, IL. 1989. pp. 21-51.
10. Damodaran, S. Structure-function relationship of food proteins. In *Protein Functionality in Food Proteins*; Hettiarchchy, N.S.; Ziegler, G.R., Eds.; Marcel Dekker, Inc. New York, NY. 1994. pp. 1-36.
11. Dev, D. K.; Quensel, E. Functional properties and microstructural characteristics of linseed flour and protein isolate. *Lebensm. Wiss. U. Technol.* **1986**, *19*, 331-337.
12. FAO/WHO. Energy and protein requirements. Report of a Joint FAO/WHO Adhoc Expert Committee. World Health Organization Technical Report Series 522. WHO, Geneva. 1973.
13. Feeney, R. E.; Whitaker, J.R. Chemical and enzymatic modification of plant proteins. In *New Protein Foods*, Vol. 5, *Seed Storage Proteins*; Altschul, A. M.; Wilcke, H. L., Eds.; Academic Press: New York, 1985; pp 181-219.

118

14. Feeney, R.E.. Yamasaki, R.B. and Geoghegan, K.F. Chemical modification of proteins. An overview. In *Modification of Proteins*; Feeney, R.E.; Whitaker, I.R., Eds.; ACS Symposium Sereis 198, American Chemical Society, Washington, DC. 1982. pp. 3-55.

15. Franzen, K. L.; Kinsella, J. E. Functional properties of acetylated and succinylated soy protein. *J. Agric. Food Chem.* **1976a**, *24*, 788-795.

16. Franzen, K. L.; Kinsella, J. E. Functional properties of succinylated and acetylated leaf proteins. *J. Agric. Food Chem.* **1976b**, *24*, 914-919.

17. Groninger, H. S. Preparation and properties of succinylated fish myofibrilla proteins. *J. Agric. Food Chem.* **1973**, *21*, 978-981.

18. Gueguen, J.; Bollecker, S.; Schwenke, K. D.; Raab, B. Effect of succinylation on some physicochemical and functional properties of the 12S storage protein from rapeseed (*Brassica napus* L.). *J. Agric. Food Chem.* **1990**, *38*, 61-69.

19. Habeeb, A. F. S. A.; Cassidy, H.; Singer, S. Molecular structural effects produced in proteins by reaction with succinic anhydride. *Biophys. Acta* **1958**, *29*, 587-597.

20. Halling, P. J. Protein stabilized foams and emulsions. *CRC Crit. Rev. Food Sci. Nutr.* **1981**, *15*, 155-203.

21. Hayakawa, S.; Nakai, S. Relationships of hydrophobicity and net charge to the solubility of milk and soy proteins. *J. Food Sci.* **1985**, *50*, 486-491.

22. Hirs, C.H.W. and Timasheff, S.N. *Methods in Enzymology*, Vol 25, Academic Press, New York, NY. 1972.

23. Johnson, E. A.; Brekke, C. J. Functional properties of acylated pea proteins isolates. *J. Food Sci.* **1983**, *48*, 722-725.

24. Kim, K. S.; Rhee, J.S. Effect of acetylation on physicochemical properties of 11S soy proteins. *J. Food Biochem.* **1989**, *13*, 187-199.

25. Kinsella, J. E. Functional properties of protein in foods; a survey. *CRC Crit. Rev. Food Sci. Nutr.* **1976**, *7*, 219-280.

26. Kinsella, J.E. Functional properties of soy proteins. *J. Am. Oil Chem. Soc.* **1979**, *56*, 242-258.

27. Kinsella, J.E. Relationships between structure and functional properties of food proteins. In *Food Proteins*; Fox, P.F.; Condon, J.E., Eds.; Applied Science Publishers, London, 1982, pp. 51-103.

28. Ma, C. -Y.; Wood, D. F. Functional properties of oat protein modified by acylation, trypsin hydrolysis and linoleate treatment. *J. Am. Oil Chem. Soc.* **1987**, *64*, 1726-1731.

29. Madhusudhan, K. T.; Singh, N. Studies on linseed protein. *J. Agric. Food Chem.* **1983**, *31*, 959-963.

30. Madhusudhan, K. T.; Singh, N. Effect of detoxification treatment on the physiochemical properties of linseed protein. *J. Agric. Food Chem.* **1985**, *33*, 1219-1222.

31. Means, G.E.; Feeney, R.E. *Chemical Modification of Proteins.* Holden-Day, San Francisco, CA, 1971.

32. Melnychym, P.; Stapley, R. Acylated protein from coffee whitener formulations. US Patent #3,764,711, **1973**.

33. Nakai, S. Structure-function relationships of food proteins with an emphasis on the importance of protein hydrophobicity. *J. Agric. Food Chem.* **1983**, *31*, 676-683.

34. Nakai, S.; Li-Chan, E. *Hydrophobic Interaction in Food Systems*, CRC Press, Boca Raton, FL., 1988.

35. Nakai, S.; Li-Chan, E.; Hirotsuka, M.; Vazquez, M. C.; Arteaga, G. Quantization of hydrophobicity for elucidating the structure-activity relationships of food proteins. In *Interaction of Food Proteins*; Parris, N.; Barford, R., Eds.; ACS Symposium Series 454, American Chemical Society, Washington, DC., 1991; pp 42-58.

36. Nitecka, E.; Raab, B.; Schwenke, K. D. Chemical modification of proteins. Part 12. Effect of succinylation on some physicochemical and functional properties of the albumin fraction from rapeseed (*B. napus* L.). *Die Nahrung* **1986**, *30*, 975-985.

37. Oomah, B. D.; Mazza, G. Flaxseed proteins - a review. *Food Chem.* **1993**, *48*, 109-114.

38. Paulson, A. T.; Tung, M. A. Solubility, hydrophobicity and net charge of succinylated canola protein isolate. *J. Food Sci.* **1987**, *52*, 1557-1561, 1567.

39. Paulson, A. T.; Tung, M. A. Emulsification properties of succinylated canola protein isolates. *J. Food Sci.* **1988a**, *53*, 817-820,825.

40. Paulson, A. T.; Tung, M. A. Rheology and microstructure of succinylated canola protein isolates. *J. Food. Sci.* **1988b**, *53*, 821-825.

41. Paulson, A. T.; Tung, M. A. Thermal induced gelation of succinylated canola protein isolate. *J. Agric. Food Chem.* **1989**, *37*, 319-328.

42. Pearce, K. N.; Kinsella, J. E. Emulsifying properties of proteins; evaluation of a turbidimetric technique. *J. Agric. Food. Chem.* **1978**, *26*, 716-723.

43. Petruccelli, S.; Añón, C. Relationship between the method of obtention and the structural and functional properties of soy protein isolates: I. structural and hydration properties. *J. Agric. Food. Chem.* **1994**, *42*, 2161-2169.

44. Petruccelli, S.; Añón, C. Relationship between the method of obtention and the structural and functional properties of soy protein isolates: I. surface properties. *J. Agric. Food. Chem.* **1994**, *42*, 2170-2176.

45. Ponnampalam, R.; Vijayalashmi, M. A.; Lenieux, L.; Amiot, J. Effect of acetylation on composition of phenolic acid and proteolysis of rapeseed flour. *J. Food Sci.* **1987**, *52*, 1552-1556, 1594.

46. Ponnampalam, R.; Delisle, J.; Gagne, Y.; Amiot, J. Functional and nutritional properties of acylated rapeseed proteins. *J. Am. Oil Chem. Soc.* **1990**, *67*, 531-536.

47. Schwenke, K. D.; Mothers, R.; Zirwer, D.; Gueguen, J.; Subirade, M. Modification of the structure of 11S globulins from plant seeds by succinylation. In *Food Proteins, Storage and Functionality*; Schwenke, K.D.; Mothers, R., Eds.; VCH Publisher, New York, 1993, pp 143-153.

48. Shukla, T. P. Chemical modification of food proteins. In *Food Protein Deterioration, Mechanism and Functionality*; Cherry, J.P., Ed.; ACS Symposium Series 206, American Chemical Society, Washington, DC. 1982, pp 275-300.

49. Sosulski, F.W. and Sarwar, G. Amino acid composition of oilseed meals and protein isolates. *Can. Inst. Food Sci. Technol. J.* **1973**, *6*, 1-5.

50. Thompson, L. U.; Cho, Y. -S. Chemical composition and functional properties of acylated low phytate rapeseed protein isolate. *J. Food Sci.* **1984a**, *49*, 1584-1587.

51. Thompson, L. U.; Cho, Y. -S. Effect of acylation upon extractability of nitrogen, phenolic acid and minerals in rapeseed flour protein concentrates. *J. Food Sci.* **1984b**, *49*, 771-776.

52. Vassel, B.; Nesbitt, L. L. The nitrogenous constituents of flaxseed II. The isolation of a purified protein fraction. *J. Biol. Chem.* **1945**, *159*, 571-584.

53. Voutsinas, L. P.; Nakai, S. A simple turbidimetric method for determining fat binding capacity of proteins. *J. Agric. Food. Chem.* **1983**, *31*, 58-63.

54. Wanasundara, P.K.J.P.D. and Shahidi, F. Functional properties and amino acid composition of solvent extracted flaxseed meal. *Food Chem.* **1994**, *49*, 45-51.

55. Wanasundara, P.K.J.P.D. and Shahidi, F. Optimization of sodium-hexametaphosphate-assisted extraction of flaxseed proteins. J. Food Sci., **1996**, *61*, 604-607.

56. Wanasundara, P. K. J. P. D.; Shahidi, F. Removal of flaxseed coat mucilage by chemical and enzymatic treatments. *Food Chem.* **1997a**, *57*, 47-55.

57. Wanasundara, P. K. J. P. D.; Shahidi, F. Functional properties of acylated flax protein isolates. *J. Agric. Food Chem.* **1997b**, *45*, 2431-2441.

58. Wang, J. C.; Kinsella, E. Functional properties of novel proteins, alfa-alfa leaf proteins. *J. Food Sci.* **1976**, *41*, 286-292.

59. Waniska, R. D.; Kinsella, J. E. Foaming properties of proteins: evaluation of a column aereation apparatus using ovalbumin. *J. Food Sci.* **1979**, *44*, 1398-1411.

60. Watanabe, M.; Arai, S. Proteinaceous surfactant prepared by covalent attachment of L-leucine n-alkyl esters to food proteins by modification with papain. In *Modification of Proteins, Food, Nutritional and pharmacological aspects*; Feeney, R.E.; Whitaker, J.R., Eds.; Advance in Chemical Series 198, American Chemical Society, Washington, DC., 1982, pp 198-222.

ANIMAL PROTEIN FUNCTIONALITIES

Chapter 8

Structure Function Relationships in Milk-Clotting Enzymes: Pepsin—A Model

R. Y. Yada and T. Tanaka

Department of Food Science, University of Guelph, Guelph, Ontario N1G 2W1, Canada

Numerous studies examining functionality on a mechanistic level of food-related proteins have recently been undertaken using genetic engineering techniques. Research in our laboratory has concentrated on the genetic engineering of pepsin, a milk-clotting enzyme which belongs to aspartic proteinases, in order to determine the role of specific amino acids/regions of pepsin, and its zymogen, pepsinogen, on structure and catalytic activity. Results from these studies indicated that site-directed mutagenesis of regions both adjacent (e.g., flap loop region) and remote (e.g., prosegment of the zymogen, surface of the enzyme) to the active site were critical to catalytic parameters and were reflected in structural changes as determined by circular dichroism and molecular modelling. Information gleaned from such studies may allow us to redesign enzymes, as well as other food-related proteins, for a particular function and/or environment in a predictable manner.

Aspartic proteinases (e.g., chymosin, pepsin, some fungal enzymes) are a class of enzymes used to clot milk in the cheese-making process and are characterized by the presence of two catalytically essential aspartic acid residues. A sequence and structural homology have been demonstrated for a number of these proteinases (1, 2), yet they often show dissimilarity in physicochemical properties such as proteolysis (e.g., chymosin, the enzyme traditionally used in the cheese-making process, has a higher milk-clotting (specific proteolysis) to general proteolytic activity ratio than pepsin), pH optimum and thermostability. The aspartic proteinases also offer a variety of protein types consisting of both glycoproteins (e.g., *Mucor miehei* proteinase) and non-conjugated proteins.

Our present research has used pepsin, and its zymogen pepsinogen, as the model proteins. Porcine pepsin (EC 3.4.23.1) was one of the first enzymes to be studied and the second to be crystallized (3). Pepsin is composed of 326 amino acid residues, has a calculated molecular weight of 35 kDa (4) and is characterized by its low milk-clotting/proteolytic (C/P) activity when compared with the milk-clotting enzyme of choice, chymosin (5). Porcine pepsinogen A, the zymogen of pepsin, is a 40 kDa protein consisting of a 44 residue prosegment that is rich in basic residues in addition to the pepsin portion which is highly acidic. The crystal structure of porcine pepsinogen A has been determined and shows that the prosegment covers and occupies the substrate binding cleft and that six of the basic residues in the prosegment are engaged in electrostatic interactions with acidic residues in the pepsin portion of the molecule.

The biological activity of a protein depends on its structure. The exact nature of this relationship has, however, remained relatively elusive. Early biochemists, enamoured with the study of structure-function relationships, gained much of today's understanding about proteins through chemical modification of protein structure (6). Chemical modification techniques, however, are compromised to various degrees by their general lack of specificity (7). The advent of molecular biology techniques has greatly aided our ability to refine these relationships by allowing for site specific alterations of the primary sequence of the protein, and has substantially enhanced our ability to study the impact of these alterations on structure-function relationships. Sites targeted by mutagenesis are frequently chosen based on information or ideas from previous research. When a protein's native structure is known, the position and mutation made are based on sound scientific hypothesis. Targets of site-directed mutagenesis often occupy the active site of enzymes. Mutations of certain highly functional residues will totally eliminate the biological activity of the protein (e.g., catalytic aspartyl groups of aspartic acid proteinase) (8). It can be desirable to modify the specificity of an enzyme by allowing secondary substrate binding in the active site. However, certain other residues governing substrate recognition could potentially be altered to either increase or decrease the specificity of the enzyme for a substrate.

Many residues (up to 50%) of the protein sequence can be mutated with no consequence to enzyme activity (9). However, mutations far from the active site can also have a perturbing effect on the conformation of the substrate recognition site thereby altering its specificity (10). Thus it is noteworthy that the net activity of a protein is a function of its entire structure, although many researchers have focused on identifying critical active site residues. Other mutations could have a dramatic global effect on the secondary and tertiary structures completely destroying functional activity. Mutations which cause disruptive global alterations to the conformation of a protein are usually avoided by selecting residues residing outside of the scaffolding structure of the protein (regions of β-sheet and α-helix) (9). Although genetic and protein engineering techniques have become prevalent in many disciplines, they are still relatively unknown in food science. Our laboratory has recently undertaken a molecular biology approach to study the structure-function relationships of aspartic proteinases, using pepsin(ogen) as the model protein, a cursory account of it is provided in this contribution (11-15). This work represents a logical extension of our previous chemical modification research (16-20) of various aspartic proteinases which yielded valuable fundamental information.

Rationale for the production of a soluble enzyme. Although the three-dimensional structure of pepsin has been successfully elucidated, only a few studies have appeared regarding putative functional residues and regions *(8, 11)*. This, in part, has been due to difficulties in expressing cloned pepsinogen (pepsin is obtained from the activation of pepsinogen) which has been in the form of inclusion bodies. Similarly, all cloned aspartic proteases using the *Escherichia coli* system, with the exception of human immunodeficiency virus protease *(21, 22)* have been expressed as inclusion bodies *(8, 23-27)*. Inclusion body proteins require that they be unfolded and then refolded in order that a "properly" folded and functional protein is obtained. Refolding, however, does not ensure that all the protein molecules have the identical and correct structure. By contrast, if the protein can be expressed in the soluble fraction and is identical in specific activity to the native protein, it is assumed that every protein molecule is folded correctly. The development of an expression system to produce soluble protein would greatly aid in studies whereby the effects of mutation(s) on structure-function of both pepsin and pepsinogen could be examined.

Fusion protein systems have been used to obtain soluble proteins *(28, 29)*. However, a common drawback to these systems is the requirement to remove the fused portion which is usually accomplished through limited proteolysis. Several proteases have been used for this purpose and include: enterokinase, thrombin, and blood coagulation factor Xa. These proteases are generally expensive, and therefore, not economical for large scale purification of the target protein. In the case of pepsin, two possible strategies to recover this protein or pepsinogen from the fusion protein without the use of expensive proteases exist: one is the autocatalytic-cleavage of fusion pepsinogen to produce pepsin, and the other is the tryptic digestion of the fused protein to recover pepsinogen. The fusion pepsinogen could cleave the fused protein autocatalytically under acidic conditions since pepsinogen excises its prosegment, which is placed between the fused protein and pepsin. Alternatively, since pepsin has only one lysine and two arginine residues in its sequence, tryptic digestion would be expected to retain the intact pepsin portion by hydrolyzing the fused protein in the linker region or in the prosegment region *(12)*.

Recently, we developed a system for the production of soluble porcine pepsinogen A (EC 3.4.23.1) by fusing the pepsinogen and thioredoxin *(29)* genes and then expressing the fused product (Trx-PG) in *Escherichia coli (12)*. The trypsin digestion of the fusion protein (to cleave off the thioredoxin) yielded recombinant pepsinogen (r-pepsinogen) whose amino terminal sequence was F-M-L-V-K-V-P-L-V-R. The sequence showed one extra residue (F) as compared to the amino terminus of the pepsinogen molecule coded in cDNA. The r-pepsinogen preparation had no milk-clotting activity prior to acidification and showed the same activity (27.7 units/mg) as commercial porcine pepsin A (n-pepsin) after acidification (Table I). The r-pepsin purified directly from the fusion protein had similar milk-clotting activity to pepsin obtained from the r-pepsinogen preparation (Table I). The amino terminal sequence of r-pepsin was I-G-D-P-E-P-L which was consistent with native porcine pepsin A *(8)*.

The milk-clotting, proteolytic activities and kinetic parameters of various pepsins are summarized in Table I. Both r- and n-pepsins showed similar milk-clotting and proteolytic activities resulting in similar milk-clotting to proteolysis activity ratios (1.41 and 1.34, respectively). Michaelis and rate constants for both pepsins were similar. Figure 1 shows the results of pH dependency of both r- and n-pepsins. Both pepsins maintained 80% activity in the range from pH 1.1 to 4.0, and almost no activity at pH 6.0.

In conclusion, it was confirmed that: 1) the fusion pepsinogen expression system produced soluble Trx-PG (a fusion protein of thioredoxin and pepsinogen); 2) r-pepsinogen and r-pepsin were similar to their respective native proteins; and 3) r-pepsin could be obtained either through direct acidification of Trx-PG or indirectly through the acidification of r-pepsinogen preparation from Trx-PG following trypsinolysis. The flexibility to produce either pepsinogen or pepsin from the fusion protein provides an excellent means by which mutation(s) of these proteins on structure-function can be examined.

Rationale for activation studies. Pepsinogen, the zymogen of pepsin, is hydrolyzed at Leu44p-Ile1 bond resulting in pepsin. This activation which occurs at acidic pH values changes the static charges of residues and is followed by the destabilization of the prosegment *(30)*. Examination of the crystal structures of pepsin and pepsinogen indicates that some drastic conformation changes occur during activation *(30-32)*. The activation mechanism of native pepsinogen into pepsin can occur via two different mechanisms. One is a bimolecular reaction (an intermolecular reaction), in which a pepsin molecule converts pepsinogen into pepsin, and the other is an unimolecular reaction (self-activation; an intramolecular reaction), in which a pepsinogen molecule cleaves itself to yield a pepsin molecule *(33-36)*. Al-Janabi et al. *(35)* in a study of pepsinogen activation, observed that below pH 4.0, in the pH range which activation occurs, both unimolecular and bimolecular activation were observed. At pH 3.0, the bimolecular activation was 6.5 times faster than self-activation while at pH 2.0 self-activation was two times faster *(35)*.

Our engineered pepsinogen (Trx-PG), which was a fusion protein of thioredoxin and pepsinogen, exhibited dominant self-activation (unimolecular reaction; intramolecular activation) in contrast to recombinant pepsinogen which exhibited both unimolecular and bimolecular reactions (intermolecular activation mediated by pepsin released during activation). At pH values of 1.1, 2.0 and 3.0, activation curves for the Trx-PG were hyperbolic rather than sigmoidal, indicating that self-activation was the dominant activation mechanism in comparison to the slower bimolecular activation (Figure 2). In order to confirm which activation mechanism was dominant, an equal mole of pepsin was added to accelerate the bimolecular reaction during activation. The addition of exogenous pepsin did not affect the activation rate of the Trx-PG but accelerated r-PG activation through the bimolecular reaction (Figure 2). In conclusion, the unimolecular activation of the fusion protein was dominant and different from that of r-PG. The difference in the activation mechanism was definitely caused by thioredoxin. The presence of thioredoxin itself, however, did not alter r-PG activation mechanism since the sigmoidal curve of r-PG activation, which was conducted in the presence of the cleaved 'prosegment' *i.e.*, thioredoxin, indicated that the unimolecular activation was extremely slow even in the presence of thioredoxin molecules. Only when thioredoxin is covalently bonded to pepsinogen was the unimolecular activation dominant. The thioredoxin portion is approximately one quarter of the entire fusion molecule (Figure 3a). The bulky 'prosegment' of the fusion protein would stabilize the

Table I. Kinetic analysis of recombinant and commercial pepsin

Abbreviation: ND, not determined.

Enzyme	Milk-clotting activity (units/mg)	Proteolytic activity (units/mg)	K_m (mM)	k_0 (s^{-1})
r-Pepsin*	27.9±1.5	19.8±0.3	0.033±0.005	65.4±3.1
r-Pepsin†	27.7±1.6	ND	ND	ND
n-Pepsin‡	28.3±0.8	21.1±0.5	0.026±0.004	79.5±3.8

* r-Pepsin purified directly from Trx-PG
† r-Pepsin purified from r-pepsinogen
‡ n-Pepsin - commercial pepsin

SOURCE: Adapted from ref. 2.

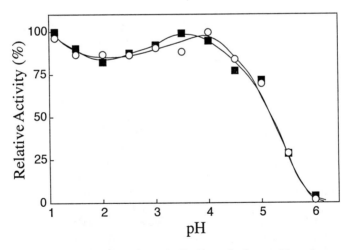

Figure 1. pH dependency of hydrolysis of synthetic peptide substrate with pepsins. Hydrolysis of the synthetic substrate was measured at the appropriate pH. Data are reported as percent relative activity, as a percentage of the highest activity over the pH range examined for the enzyme in question. Open circles and closed boxes represent r-pepsin and n-pepsin, respectively. (Adapted from ref. 2).

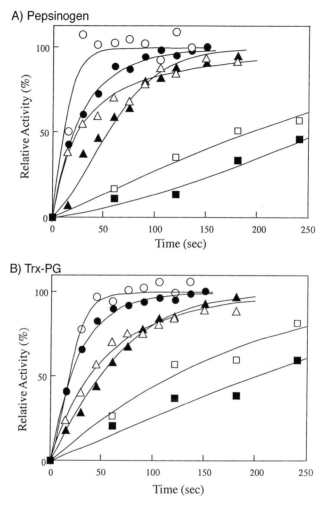

Figure 2. Activation time course of recombinant pepsinogen (r-PG, A) and fusion pepsinogen (Trx-PG, B). The ratio of activated zymogens was calculated from the activity measurements. Nonactivated samples are defined as 0% and the plateau of the activation curves is defined as 100%. Open symbols represent pepsinogen activated in the presence of exogenous pepsin, while solid symbols represent pepsinogen activated in the absence of exogenous pepsin. Circles, triangles, and squares represent pH 1.1, 2.0 and 3.0, respectively. Each data point represents the mean of a minimum of three determinations. (Adapted from ref. 3).

128

a)

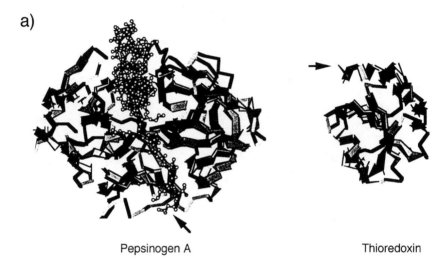

Pepsinogen A   Thioredoxin

Figure 3.   Three-dimensional structure of pepsinogen and thioredoxin, and schematic scheme of activation of Trx-PG. (a) Three-dimensional structures of pepsinogen and thioredoxin (Protein Data Bank 3PSG and 2TRX, respectively) are shown in the same scale. The prosegment of pepsinogen is shown in ball-and-stick models. The N-terminal of pepsinogen and C-terminal of thioredoxin are indicated with arrows. Thioredoxin consists of 108 amino acid residues while pepsinogen is 371 amino acid residue long.   In Trx-PG, both protein are connected by a 20-amino-acid residue linker.   (b) Proposed scheme of how thioredoxin prevents pepsin from cleaving the fusion protein. In both the fusion protein (Trx-PG) and pepsinogen, two aspartic active sites (two "D"s in the figure) are covered with a prosegment (the thick lines in the figure). Pepsinogen can be activated into pepsin through either self-activation or bimolecular activation. Trx-PG has a large independent domain, i.e., thioredoxin portion, at the amino terminal of the prosegment. This large domain would prevent pepsin from approaching the susceptible site on the prosegment, and therefore, the bimolecular reaction could not occur. (Adapted from ref. 3).

b)

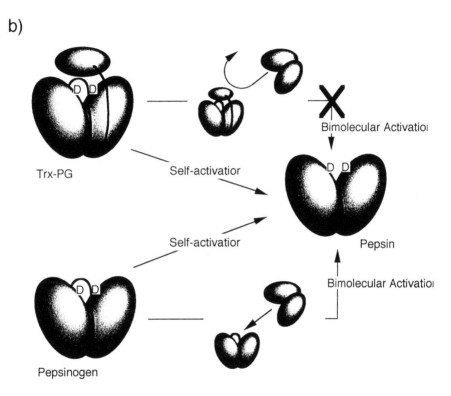

Figure 3. *Continued.*

catalytic intermediate of unimolecular activation and could also retard bimolecular activation. The bulkiness of the 'prosegment' of the fusion protein could prevent pepsin molecules from approaching the cleavage site, Leu44p-Ile1 (Figure 3b). The nature of the fusion protein could provide an effective means to estimate the effect of mutations on the activation of pepsinogen, e.g., mutations in the prosegment.

Rationale for studying residues distal to the active site. Previous investigations *(8, 37)* into the functional activity of pepsin have been largely restricted to residues of the substrate binding pocket/active site due to their importance. This approach, however, assumes that the bulk of the protein structure plays a minor role in the functional activity of the enzyme. It has been suggested, however, that enzymes behave more like mechanical devices than static bio-catalysts implying areas external to the active site are vital to enzymatic activity *(10)*.

Lys319 is the sole lysine in the entire mature pepsin molecule. Located near the back side of the active cleft, Lys319 is believed to have little interaction with prosegment residues during folding and is exposed to solvent in the mature enzyme *(4)*.

In order to study the effects of mutations distal to the active site of pepsin, a study was conducted in which site-directed mutagenesis of Lys319 to methionine (a hydrophobic residue) and glutamic acid (a negatively charged residue) were generated (1). The mutations, Lys-319 →Met and Lys-319 →Glu, resulted in a progressive increase in the K_m and similar decrease in k_{cat} (Table II), respectively, as well as being denatured at a lower pH than the wild-type pepsin (Figure 4). CD analysis (Table III) indicated that mutations at Lys-319 resulted in changes in secondary structure fractions which were reflected in changes in enzymatic activity as compared to the wild-type pepsin, *i.e.* kinetic data (Table II) and pH denaturation study (Figure 4). Molecular modelling of mutant enzymes indicated differences in flexibility in the flap loop region of the active site, the region around the entrance of the active cleft, subsite regions for peptide binding, and in the subdomains of the C-terminal domain when compared to the wild-type enzyme (Figure 5). It has been proposed that certain enzymes are mechanical devices rather than static biocatalysts, with similarity to a pump in processing substrate *(10)*. This thought gives rationality to the bulk of an enzyme's structure, which is mostly thought of as incidental when considering merely the active site and subsite binding pockets, as the catalytic machinery. Pepsin has been shown to be extremely flexible in several areas including subdomains in C-terminal domain *(32)*, the entrance region of the active cleft *(38)*, and the region joining the N- and C-terminal domains. We concluded from this study that the induced mutations in pepsin at position 319, which is distal to the active site, disrupted the enzymatic activity by altering the flexibility of the protein, and thereby altering the catalytic activity of the enzyme.

Rationale of prosegment research. Synthesis of proteolytic enzymes, including aspartic proteinases, as inactive precursors is one way of regulating proteolytic activity in order to prevent undesirable degradation of proteins. These inactive precursors, known as zymogens, differ from the corresponding active enzymes by the presence of

Table II. Kinetic parameters of wild-type pepsin and mutated pepsins, Lys-319 → Glu and Lys-319 → Met

Enzyme	K_m	k_{cat}	k_{cat}/K_m
	(mM)	*(s⁻¹)*	*(mM⁻¹s⁻¹)*
Wild-type pepsin	0.035	328	9371
Lys-319 → Glu pepsin	0.246	52	211
Lys-319 → Met pepsin	0.136	215	1581

SOURCE: Adapted from ref. 1.

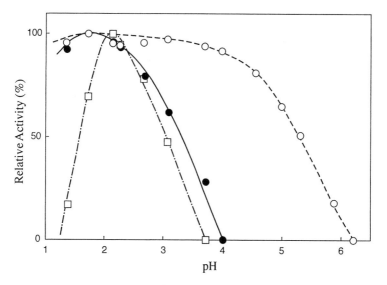

Figure 4. Activity profiles at varying pH using the synthetic substrate Lys-Pro-Ala-Glu-Phe-Phe(NO₂)-Ala-Leu, for wild-type pepsin (○), Lys-319 → Met-pepsin (●), and Lys-319 → Glu-pepsin (□). (Adapted from ref. 1).

Table III. Secondary structure fractions for wild-type pepsin and the two mutants Lys-319 → Glu and Lys-319 → Met at pH 2.1 and 5.3

Protein sample	Secondary structure				
	α-Helix	β-Sheet	β-Turn	Ramdom Coil	
			%		
Wild-type pepsin, pH 2.1	0.7	54.5	12.5	32.2	
Lys-319 → Glu pepsin, pH 2.1	17.6	25.6	18.0	38.7	
Lys-319 → Met pepsin, pH 2.1	11.0	32.6	20.2	36.2	
Wild-type pepsin, pH 5.3	3.4	63.1	7.3	26.2	
Lys-319 → Glu pepsin, pH 5.3	3.8	57.3	9.0	29.9	
Lys-319 → Met pepsin, pH 5.3	2.1	57.2	9.7	31.0	

SOURCE: Adapted from ref. 1.

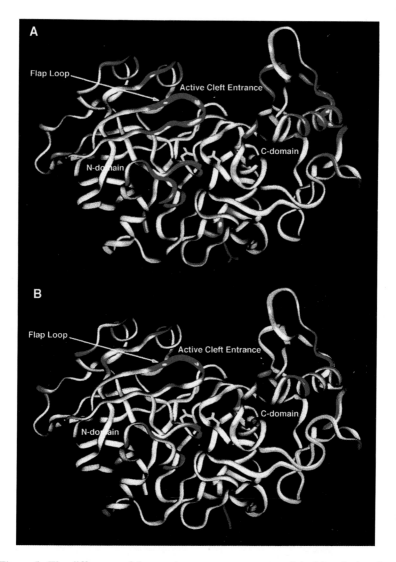

Figure 5. The difference of the maximum movement at each residue during the molecular dynamics simulations. The ribbon diagram of pepsin models show the difference between wild-type and each mutant pepsins, Lys-319 → Glu (A) and Lys-319 → Met (B). Red portions indicate the maximum movement (flexibility index) are larger that the counterpart of wild-type. Skyblue portions indicate the opposite. Grey portions mean both wild-type and mutant have similar indices. Thicker colored portions represent larger differences. Active Asp-32 (left) and -215 are in yellow, and the Glu and Met mutants are in purple. (Adapted from ref. 1).

134

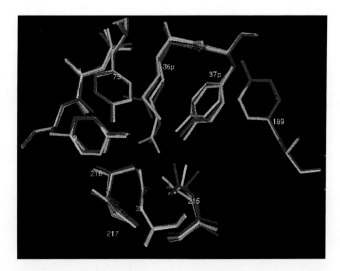

Figure 6. Superpostiion of active site residues from the molecular models of wild-type K36pR, K36pM and K36pE pepsinogen. Wild-type, purple; K36pR, yellow; K36pM, blue; K36pE, red. (Adapted from ref. 4).

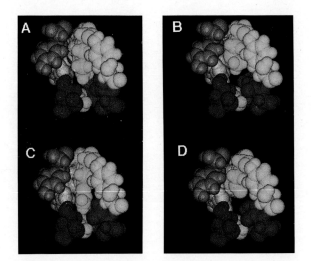

Figure 7. Space-filling diagrams of the active site of wild-type, K36pR, K36pM and K36pE pepsinogen showing the packing of residue 36p and the catalytic aspartates. Asp32, red (left); Asp215, red (right); Lys36p, dark green; Tyr37p, light green; Gly217, blue; Tyr9, yellow; Thr218, purple; Asn8, white; Tyr75, orange. Wild-type, upper left; K36pR, lower left; K36pM, upper right; K36pE, lower right. (Adapted from ref. 4).

additional amino acids (referred to as the prosegment) at the N-terminal end of the protein. Nearly all aspartic proteinases (EC 3.4.23.X) are synthesized as zymogens that are stable at neutral pH and are converted to active enzymes at acidic pH by proteolytic cleavage of the prosegment *(39, 40)*. A classic example of an aspartic proteinase zymogen is porcine pepsinogen A, the inactive precursor of pepsin, which is the enzyme responsible for digestion of dietary protein in the gastric lumen. Based on these observations, it has been proposed that at neutral pH, the prosegment maintains pepsinogen in its inactive form by blocking access to the active site and stabilizes the conformation of pepsinogen by neutralizing negative charges in the pepsin portion of the molecule *(1, 30, 41)*.

Early chemical modification studies indicated that basic residues in the prosegment of pepsinogen are critical to the conformation of pepsinogen, as well as its conversion to pepsin *(42-45)*. One of these basic residues is Lys36p, which is highly conserved in the prosegments of aspartic proteinases *(39, 40)* and in the crystal structures of porcine pepsinogen *(1, 30)* and human progastricsin *(46)* forms ion pairs with the two active site aspartates, Asp32 and Asp215, that are required for catalysis. It has been proposed that Lys36p is critical to the stability and inactivity of the zymogen at neutral pH *(1, 30)* and may be required for correct folding of the zymogen *(39)*. With respect to the latter, a basic residue corresponding to Lys36p in the aspartic proteinases yeast proteinase A *(47)* and aspergillopepsinogen I *(48)* has been reported to be required for protein folding. However, in these studies, the structure of the Lys36p mutants was not directly examined. Rather, the potential activity of the zymogen was used as a measure of protein folding and a decrease or loss in potential activity was assumed to be indicative of incorrect folding.

Given the proposed importance of Lys36p and the paucity of research on the role of this residue in the aspartic proteinases, a study was undertaken to examine the contribution of Lys36p to the structure and function of porcine pepsinogen and its active form pepsin *(15)*. This was accomplished by replacing Lys36p with the residues Arg, Met and Glu, and then expressing the mutant pepsinogens using the soluble expression system *(12)*. The effect of these mutations on the kinetic parameters of pepsin, and on the structure and activation of pepsinogen was evaluated *(15)*.

All the three mutants were less stable than the wild-type pepsinogen. During purification, all the mutants began to degrad while the wild-type was stable under the same conditions. Especially noteworthy was K36pE where it was impossible to measure the activation rate. Activation kinetics indicated that K36pR and K36pM were activated faster than wild-type (Table IV). While the positive charged replacement, K36pR, showed up to three times the rate as wild-type, the neutral replacement, K36pM, accelerated up to ten times the rate. If it is assumed that the instability of K36pE was due to easier displacement of the prosegment, the results indicated the positive charge on residue 36p had a key role in maintaining the stable zymogen. Molecular modelling suggested that mutating Lys36p resulted in the molecule assuming a different conformation of pepsinogen (Figures 6 and 7). These structural analyses showed the different conformations of the prosegment, in which the position of active site residues, such as Asp215, were relocated. These results not only supported the faster activation rates but also suggested different conformation of activated pepsin.

The kinetic constants of pepsins derived from Lys36p mutants showed similar K_m values but slower k_0 values than wild-type for two different substrates (Table V). From the results, it was concluded that 1) Lys36p is necessary to stabilize the conformation of the zymogen, and 2) Lys36p is accordingly essential to determine the position of the active site residues to achieve high activity of pepsin.

Rationale of flap loop region. Many enzymes have flexible loops, floppy tails, mobile lids or moving hinged domains which appear to specifically contribute to the mechanism of enzyme catalysis *(49)*. Structural studies on enzyme/inhibitor complexes with fungal aspartic proteinases provided the first insights into substrate binding and on the ligand-induced structural changes. Initial X-ray diffraction studies have shown that there is a very flexible β-hairpin flap located at the entrance of the active site (residues 71 - 82). The flap ensures an environment secluded from the solvent such that transient substrate intermediates are encapsulated by the flap and stabilized *(50-54)*. Three flap residues contribute directly to substrate subsite specificity. Threonine (Thr) 77 contributes to S_1 and S_2 subsites, Glycine (Gly) 76 to S_2 subsite and Tyr75 to S_1 and S_2' subsite (41). Tyr75 is conserved in all eukaryotic aspartic proteinases and is located near the tip of the flap and its phenolic ring forms one of the walls of the S_1 substrate-binding pocket *(50, 51, 55)*. The position of Tyr75, which is determined through hydrogen bonding with Trp39, has been suggested to stabilize the transition state *(56)*. Tang and Koelsch *(57)* proposed a role of Tyr75 for substrate capture as well as aligning the substrate for cleavage. A study was undertaken in which Tyr75 was mutated to Phe and Asn (Phe lacks a *p*-hydroxyl group and Asn resembles of the bottom of the phenolic ring) *(14)*.

The pH dependency among WT-, Y75F- and Y75N-pepsin was similar (Figure 8). If the position of the loop had been affected with the replacement, it would have affected the pH dependency since the loop in wild type protects the active site from effects of the outside environment *(54)*. The results, therefore, suggested that the mutation of Tyr75 did not affect the loop's ability to maintain an environment favorable for reaction.

Despite minor substantial changes in far UV circular dichroism spectra (data not shown) and pH dependency, the substrate preference was altered. The ratios of milk-clotting activity to proteolytic activity (MC/PA ratio) varied among the Y75F, Y75N, and WT pepsins, *i.e.*, 1.7, 5.7, and 2.9, respectively (Table VI). These changes suggested that the hydrophobic pocket which contributes to the specificity of pepsin for the substrate P1 site had been altered. However, since WT-pepsin has a broader substrate specificity compared to chymosin *(58, 59)* and since Y75N-pepsin still forms the hydrophobic pocket with other residues, e.g., Phe111 and Leu120, it was postulated that the absence of the ring structure did not greatly affect specificity as compared to chymosin Y75N mutant *(27)*, *i.e.*, the altered pocket would help in the alignment of κ-casein as well as nonspecific peptides. Y75F mutant retained the aromatic ring to form the pocket, and, accordingly, did not change the substrate preference.

Kinetic parameters, determined by the hydrolysis of the synthetic oligopeptides are summarized in Table VII *i.e.*, Lys-Pro-Ala-Glu-Phe-Phe (NO₂)-Ala-Leu (Peptide 1, Phe (NO₂) is p-nitrophenylalanine); Leu-Ser-Phe (NO₂)-Nle-Ala-Leu methyl ester (Peptide

Table IV. Kinetics constants of pepsin from Trx-PG mutants

KPAEFF(NO$_2$)AL Pepsin	K_m (mM)	k_o (s^{-1})
Wild-type	0.033 ± 0.005	65.4 ± 3.1
K36pR	0.025 ± 0.003	31.3 ± 1.0
K36pM	0.024 ± 0.003	6.2 ± 0.2
K36pE	0.017 ± 0.004	4.2 ± 0.2
LSF(NO$_2$)NleAL-OCH$_3$ Pepsin	K_m (mM)	k_o (min^{-1})
Wild-type	0.019 ± 0.006	190.5 ± 18.8
K36pR	0.011 ± 0.003	11.6 ± 1.2
K36pM	0.034 ± 0.003	0.25 ± 0.01
K36pE	0.024 ± 0.011	0.036 ± 0.006

SOURCE: Adapted from ref. 5.

Table V. Activation rate constants of K36p mutants

pH	First-order rate constant (min^{-1}) ± SD		
	wild-type	K36pR	K36pM
1.1	1.10 ± 0.10	5.80 ± 0.51	10.3 ± 1.2
2.0	0.519 ± 0.058	1.25 ± 0.13	6.56 ± 0.64
3.0	0.109 ± 0.011	0.185 ± 0.025	0.675 ± 0.012

SOURCE: Adapted from ref. 5.

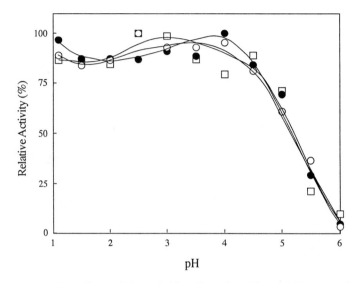

Figure 8. pH dependency of the activities of pepsins. The activities at variety of pH were determined with using synthetic oligopeptide substrates. WT (●), Y75N (○) and Y75F (□). (Adapted from ref. 5).

Table VI. Comparison of milk-clotting and proteolytic activities

Pepsin	Milk-clotting (unit)	Proteolysis (unit)	MC/PA
Wild-type	1.56±0.03	0.54±0.00	2.9
Y75F	1.54±0.03	0.89±0.01	1.7
Y75N	0.86±0.07	0.15±0.01	5.7

MC = milk-clotting
PA = proteolytic activity

SOURCE: Adapted from ref. 4.

Table VII. Kinetic constants for synthetic substrates

Pepsin	Peptide 1 (pH 2.1)		Peptide 2 (pH 3.95)	
	K_m (mM)	k_0 (S^{-1})	K_m (mM)	k_0 (min^{-1})
Wild-type	0.034±0.005	65.4±3.1	0.019±0.006	190.5±18.9
Y75F	0.026±0.003	82.6±3.3	0.016±0.007	70.5±13.7
Y75N	0.036±0.006	12.2±0.7	0.012±0.003	25.6±1.6

Peptide 1 = Lys-Pro-Ala-Glu-Phe-Phe(NO_2)-Ala-Leu, Phe(NO_2) is
p-nitrophenylalanine
Peptide 2 = Leu-Ser-Phe (NO_2)-Nle-Ala-Leu methyl ester

SOURCE: Adapted from ref. 4.

2). Michaelis constants of either substrate for wild-type and mutant pepsins were similar within error values although slower catalytic rates were observed for mutants. A larger change of Michaelis constants with substitution of Tyr75 was expected since it forms a wall of the hydrophobic pocket of the active site. However, the results showed small differences among WT and mutant porcine pepsins for both substrate while results observed in *Rhizomucor* pepsin *(60, 61)* and chymosin *(62)* varied among the substrate, reflecting the preferences of substrate, *i.e.*, small change for a preferred substrate. Since both substrates are good substrates for pepsin, it was not certain how Tyr75 contributes to substrate specificity, given that the results indicated that the tyrosine residue was not critical for substrate binding.

The results of the other aspartic proteases *(60-62)* showed a similar tendency that Tyr75 substitution generally decrease the reaction rate. However, the importance of the aromatic ring varies among the aspartic proteases. The most comprehensive mutation work on Tyr75, reported on *Rhizomucor pusillus* pepsin *(60, 61)*, showed the absence of the aromatic ring enhanced the catalytic efficiency by 5 times on the substrate, Leu-Ser-Phe(NO_2)-Nle-Ala-Leu methyl ester (Peptide 2). With the same substrate, our study showed slower catalysis with Asn substitution as well as Phe substitution. Depending on the substitution, the reaction rates decreased from 2 to 4 times.

Since the pH dependency (Figure 8) showed no differences among pepsins, it was, therefore, assumed that the environment did not vary among the pepsins. It was postulated that the substitution altered the non covalent bonding networks which keep the active site in a reactive conformation, therefore, differences in distances between the inhibitor and the substrate binding site among the pepsins would be observed. Molecular modelling allowed for evaluation of these distances (Table VIII). WT and Y75F pepsins showed that the average distance between the cleavage site (an oxygen atom O2 of I4 site) and both active aspartates (Asp32 and 215) was less than 3.7Å, whereas the average distance between Asp32 and the cleavage site exceeded 4.1 Å for Y75N pepsin and was 6.0Å at the furthest point. Since the theoretical distance between two hydrogen-bonding oxygen atoms is 2.76Å, the aspartic residue of Y75N would not be able to readily interact with the inhibitor. Therefore, the catalytic aspartate may not properly function during catalysis.

From the above results, it was concluded that 1) Tyr75 aligns the substrate in a favourable position for reaction, 2) Tyr75 had less impact on the specificity of the substrates as compared to chymosin and *Rhizomucor* pepsin which have higher substrate specificity, and 3) neither the positioning of the prosegment nor closure of the loop was affected by mutations at Tyr75. Since porcine pepsin has broader specificity as compared to the other aspartic proteases, the hydrophobic pocket which is thought to determine a part of substrate specificity, *i.e.*, P1 site, may play a less important role in catalysis.

Conclusion

Recently, studies examining the molecular basis of food protein functionality have become more prevalent due largely to the use of genetic engineering techniques such as site-directed mutagenesis. In undertaking our recent research into the structure-function relationship of pepsin(ogen), it was our intent to provide fundamental

Table VIII. The distance between pepsins and inhibitor during the dynamics calculation

Inhibitor	Pepsin	WT			Y75F			Y75N		
		max	min	avg	max	min	avg	max	min	avg
I2:HN	219:OG	6.1	1.8	3.2	8.9	2.0	5.7	4.2	1.8	2.4
I2:O	219:HN	5.1	1.7	3.3	6.4	1.9	4.4	3.8	1.9	2.6
I3:HN	77:OG1	6.0	1.9	3.4	3.8	1.8	2.6	3.8	1.8	2.5
I3:O	77:HN	4.3	1.8	3.0	5.0	1.8	2.9	4.2	1.7	2.5
I4:HN	217:O	4.6	2.0	3.4	5.5	2.2	4.0	4.6	2.2	3.2
I3:O	77:OG1	6.0	2.6	3.9	4.5	2.5	3.3	3.9	2.6	3.0
I3:O	76:HN	5.5	2.0	3.6	7.0	2.5	5.2	4.7	2.4	3.5
I4:O3	76:HN	7.8	2.3	5.7	9.9	2.4	7.8	4.4	1.9	2.6
I4:O2	32:OD1	4.6	2.8	3.5	5.0	2.8	3.7	5.5	2.8	4.3
I4:O2	32:OD2	4.9	2.4	3.1	4.9	2.5	3.2	6.0	2.7	4.1
I4:O2	215:OD1	4.9	2.7	3.7	4.3	2.7	3.4	4.7	2.8	3.7
I4:O2	215:OD2	5.0	2.5	3.3	4.5	2.5	3.0	4.2	2.5	3.1

information regarding mechanisms by which aspartic proteinases function. Such information may provide an effective way to match the physicochemical properties of an enzyme and the desired application through predictability rather than empiricism, as has been relied on in the past. In addition, information gleaned from structure-function relationships of pepsin(ogen) will contribute to structure function relationships of other protein systems, thereby allowing for more efficient utilization of existing food protein sources.

Acknowledgements

The financial assistance of the Natural Sciences and Engineering Research Council of Canada in supporting the above research is gratefully acknowledged. The skilful assistance of Ms. Sue Hall in the preparation of this chapter is gratefully acknowledged.

Literature Cited

1. Hartsuck, J.A., Koelsch, G., Remington, S.J. *Proteins: Struct. Funct. Genet.* **1992**, *13*, 1-25.
2. Szecsi, P.B. *Scand. J. Clin. Lab. Invest.* **1992**, *52*, 5-22.
3. Northrop, J.H. *J. Gen. Physiol.* **1930**, 13, 739.
4. Cooper, J.B., Khan, G., Taylor, G., Tickle, I.J., Blundell, T.L. *J. Mol. Biol.* **1990**, *214*, 199-222.
5. Suzuki, J., Hamu, A., Nishiyama, M., Horinouchi, S. Beppu, T. *Protein Eng.* **1990**, *4*, 69-71.
6. Schultz, G.E. *Nature* **1974**, *250*, 140-141.
7. Howell, N.K. In *Food Proteins: Properties and Characterization*; Editor, Nakai, S. and Modler, H.W. VCH Publishers, New York. **1996**. pp. 235-280.
8. Lin, X., Wong, R.N.S., Tang, J. *J. Biol. Chem.* **1989**, *264*, 4482-4489.
9. Sambrook, J., Fritsch, E.F., Maniatis, T. *Molecular Cloning: A Laboratory Manual*; Cold Spring Harbor Laboratory, Cold Spring Habor, New York, **1989**.
10. Williams, R.P.J. *T.I.B.S.* **1993**, *18*, 115-117.
11. Cottrell, T.J., Harris, L.J., Tanaka, T., Yada, R.Y. *J. Biol. Chem.* **1995**, *270*, 19974-19978.
12. Tanaka, T., Yada, R.Y. *Biochem. J.* **1996**, *315*, 443-446.
13. Tanaka, T., Yada, R.Y. *Arch. Biochem. Biophys.* **1997**, *340*, 355-358.
14. Tanaka, T., Teo, K.S.L., Harris, L.J., Yada, R.Y. *Protein and Peptide Lett.* **1998**, *85*, 19-26.
15. Richter, C., Tanaka, T., Koseki, T., Yada, R.Y. *Protein Eng.* **1997**, Submitted Dec. 1997.
16. Brown, E.D., Yada, R.Y. *Biochim. Biophys. Acta.* **1991**, *1076*, 406-415.
17. Smith, J.L., Yada, R.Y. *Agric. Biol. Chem.* **1991**, 55, 2017-2024.
18. Smith, J.L., Yada, R.Y. *J. Food Biochem.* **1991**, *15*, 331-346.
19. Smith, J.L., Billings, G.E., Yada, R.Y. *Agric. Biol. Chem.* **1991**, *55*, 2007-2016.
20. Smith, J.L., Billings, G.E., Marcone, M.F., Yada, R.Y. *J. Agric. Food Chem.* **1992**, *40*, 3-8.
21. Seelmeier, S., Schmidt, H., Turk, V., von der Helm, K. *Proc. Natl. Acad. Sci.* **1988**, *85*, 6612-6616.

22. Darke, P.L., Leu, C.-T., Davis, L.J., Heimbach, J.C., Diehl, R.E., Hill, W.S., Dixon, R.A.F., Sigal, I.S. *J. Biol. Chem.* **1989**, *264*, 2307-2312.
23. Nishimori, K., Kawaguchi, Y., Hidaka, M., Uozumi, T., Beppu, T. *Gene* **1982**, *19*, 337-344.
24. Kaytes, P.S., Theriault, N.Y., Poorman, R.A., Murakami, K., Tomich, C.-S.C. *J. Biotech.* **1986**, *4*, 205-218.
25. Masuda, T., Nakano, E., Hirose, S., Murakami, K. *Agric. Biol. Chem.* **1986**, *50*, 271-279.
26. Imai, T., Cho, T., Takamatsu, H., Hori, H., Saito, M., Masuda, T., Hirose, S., Murakami, K. *J. Biochem.* (Tokyo) **1986**, *100*, 425-432.
27. Chen, Z., Koelsch, G., Han, H.-p., Wang, Z.-J., Lin, X.-l., Hartsuck, J. A., Tang, J. *J. Biol.Chem,* **1991**, *266*, 11718-11725.
28. Nilsson, B., Abrahmsen, L., Uhlén, M. *EMBO J.* **1985**, *4*, 1075-1080.
29. LaVallie, E.R., DiBlasio, E.A., Kovacic, S., Grant, K.L., Schendel, P.F., McCoy, J.M. *Bio/Technology* **1993**, *11*, 187-193.
30. James, M.N.G., Sielecki, A.R. *Nature* **1986**, *319*, 33-38.
31. Andreeva, N.S., Zdanov, A.S., Gustchina, A.E., Fedorov, A.A. *J. Biol. Chem.* **1984**, *259*, 11353-11365.
32. Abad-Zapatero, C., Rydel, T.J., Erickson, J. *Proteins* **1990**, *8*, 62-81.
33. Herriott, R.M. *J. General Physiol.* **1939**, *22*, 65-78.
34. Bustin, M., Conway-Jacobs, A. *J. Biol. Chem.* **1971**, *246*, 615-620
35. Al-Janabi, J., Hartsuck, J.A., Tang, J. *J. Biol. Chem.* **1972**, *247*, 4628-4632
36. McPhie, P. *Biochem. Biophys. Res. Comm.*. **1974**, *56*, 789-792
37. Pearl, L.H., Bundell, T.L. *FEBS Lett.* **1964**, *174*, 96-101.
38. Chen, L. Erickson, J.W., Rydel, T.J., Park, C.H., Neidhart, D., Luly, J., Abad-Zapatero, C. *Acta Cryst.* **1992**, *B48*, 476.
39. Foltmann, B. *Biol. Chem. Hoppe-Seyler* **1988**, *369 Suppl.*, 311-314.
40. Koelsch, G., Mares, M., Metcalf, P., Fusek, M. *FEBS Lett.* **1994**, *343*, 6-10.
41. Sielecki, A.R., Fujinaga, M., Read, R.J., James, M.N.G. *J. Mol. Biol.* **1991**, *219*, 671-692.
42. Gounaris, A.D., Perlmann, G.E. *J. Biol. Chem.* **1991**, *242*, 2739-2745.
43. Rimon, S., Perlmann, G.E. *J. Biol. Chem.* **1968**, *243*, 3566-3572.
44. Nakagawa, Y., Perlmann, G.E. *Arch. Biochem. Biophys.* **1972**, *149*, 476-483.
45. Dykes, C.W., Kay, J. *Trans. Biochem. Soc.* **1977**, *5*, 1535-1537.
46. Moore, S.A., Sielecki, A.R., Chernaia, M.M, Tarasova, N.I., James, M.N.G. *J. Mol. Biol.* **1995**, *247*, 466-485.
47. van den Hazel, H.B., Kielland-Brandt, M.C., Winther, J.R. *J. Biol. Chem.* **1995**, *270*, 8602-8609.
48. Inoue, H., Hayashi, T., Huang, X., Lu, J., Athauda, S.B.P., Kong, K., Yamagata, H., Udaka, S., Takahashi, K. *Eur. J. Biochem.* **1996**, 237, 719-725.
49. Knowles, J.R. *Nature,* **1991**, *350*, 121-124.
50. Sali, A., Veerapandian, B., Cooper, J.B., Foundling, S.I., Hoover, D.J., Blundell, T.L. *EMBO J.* **1989**, *8*, 2179-2188.
51. Bott, R., Subramanian, E., Davies, D.R. *Biochemistry* **1982**, *21*, 6956-6962.
52. James, M.N.G., Sielecki, A., Salituro, F., Rich, D.H., Hofmann, T. *Proc. Natl. Acad. Sci. USA* **1982**, *79*, 6137-6146.

144

53. Abad-Zapatero, C., Rydel, T.J., Niedhart, D.J., July, J., Erickson, J.W. In *Strucuture and Function of the Aspartic Proteinases*. Editor, Dunn, B. Plenum Press, New York, **1991**. pp. 9-21.

54. Foundling, S.I. Cooper, J., Watson, F.E., Cleasby, A., Pearl, L.H., Sibanda, B.L., Hemmings, A., Wood, S.P., Blundell, T.L., Valler, T.L., Norey, C.G., Kay, J., Boger, J., Dunn, B.M., Leckie, B.J., Jones, D.M., Atrash, B., Hallett, A., Szelke, M. *Nature*, **1987**, *327*, 349-352.

55. Suguna, K., Padlan, E.A., Smith, C.W., Carson, W.D., Davies, D.R. *Proc. Natl. Acad. Sci. USA* **1987**, *84*, 7009-7013.

56. Blundell, T.L., Cooper, J., Foundling, S.I., Jones, D.M., Atrash, B., Szelke, M. *Biochemistry* **1987**, *26*, 5585-5590.

57. Tang, J., Koelsch, G. *Protein Peptide Lett.* **1995**, *2*, 257-266.

58. Baker, L.E. *J. Biol. Chem.* **1951**, *193*, 809-819.

59. Fruton, J.S. *Adv. Enzymol.* **1970**, *33*, 401-443.

60. Beppu, T., Park, Y.-N., Aikawa, J., Nishiyama, M., Horinouchi, S. In *Aspartic Proteinases*. Editor, Takahashi, K. **1995**, pp. 501-509.

61. Park, Y.-N., Aikawa, J.-i., Nishiyama, M., Horinouchi, S., Beppu, T. *Protein Eng.* **1996**, *9*, 869-875.

62. Suzuki, J., Sasaki, K., Sasao, Y., Hama, A., Kawasaki, H., Nishiyama, M., Horinouchi, S., Beppu, T. *Protein Eng.* **1989**, *2*, 563-569.

Chapter 9

Functional Properties of Whey Proteins in Forming Networks

E. A. Foegeding[1], E. A. Gwartney[1], and A. D. Errington[1,2]

Department of Food Science, North Carolina State University,
Raleigh, NC 27695–7624

Gel formation is one of the most important functionalities of whey proteins. The network structures formed in protein gelation contribute to the appearance, texture, water-holding and flavor delivery of the gel. Gel networks are classified as fine-stranded, particulate or mixed structures. In general, fine-stranded gels are transparent or translucent and have high water-holding properties, whereas particulate gels are opaque and have lower water-holding properties. Fine-stranded gels are formed when aggregation produces network strands with diameters representing one to several molecular diameters. Particulate gels are formed when there is aggregation prior to denaturation, or the aggregation of denatured molecules produces dispersed protein clusters with diameters 100-1000 times molecular diameters. Particulate gel networks appear to be formed from secondary aggregation of these clusters. The type of gel network formed is determined by pH, solutes and kinetics of gelation.

Protein Functionality

Protein functionality describes the general processes regarding how proteins impact the quality and stability of foods. The simplest, most inclusive, definition of protein functionality is the collection of protein reactions which produce a desired effect when added to, or part of, a food product. Protein functionality is related to reactions occurring in solutions (e.g., phase stability of infant formula), at interfaces (e.g., emulsion and foam stabilization) and in forming networks. The process of network formation is required to produce a protein film or gel. The formation of whey protein gel networks will be the subject of this chapter.

[2]Current address: Cryovac, Duncan, SC 29334

Mechanisms Of Whey Protein Gelation

There are two main steps involved in all gelation processes. The first is a change in the system which increases inter-molecular interactions. This can be considered the initiation or "molecular activation" phase of the overall reaction. The second step is aggregation of molecules into a gel network. As noted by Ferry (1), forming a gel network requires the proper balance between attractive and repulsive forces. An excess in attractive forces produces a precipitate whereas too many repulsive forces produces soluble aggregates.

The most common mechanism used to form whey protein gels is heat-induced protein denaturation. It is also possible to form gels by: i) denaturing with a chaotropic compound such as urea (2); ii) linking molecules covalently by enzymatic activity (e.g., transglutaminase) (3); or iii) limited proteolytic activity (4).

Ferry (1) postulated that gelation of denatured proteins can be represented as the following two-stage process:

Native protein $\rightarrow$ Denatured protein (long chains) $\rightarrow$ Network

This overall view is still valid; however, there have been some modifications in the intervening 49 years. In heat-induced gelation the extent of unfolding appears minimal because networks have been observed which are formed by strands with diameters very close to molecular diameters (5), therefore unfolding into "long chains" is not a requirement. There are also situations where soluble aggregates are formed as intermediates between the denatured protein and network stage (6).

Figure 1 shows a generalized model for heat-induced gelation of whey proteins based on one proposed by Ziegler and Foegeding (7). In all cases the reaction is started by altering protein structure. This may be loss of tertiary and secondary structure or simply the loss of tertiary structure, as in forming a molten globule state. Under the proposed gelation model a gel cannot be formed from aggregation of proteins in the molten globule state. When β-lactoglobulin is heated at 40°C, in the presence of 20 mM $CaCl_2$, aggregates are formed which do not produce a gel matrix (unpublished data). However, Hirose (8) has proposed a gelation mechanism for ovotransferrin and serum albumin based on forming a molten globule state at 37°C and introducing intermolecular disulfide bonds by sulfhydryl-disulfide interchange.

Aggregation of denatured proteins continues until a gel point is reached (Figure 1). At the gel point the molecular weight of the aggregate(s) diverge to infinite size and a gel matrix is formed. Right before the gel point the fluid viscosity is infinitely large, and right after the gel point the rigidity is infinitely small (9). The rigidity increase observed after the gel matrix is formed is due to addition of more molecules to the matrix and/or transformations within the structure of the gel matrix (e.g., formation of additional bonds among molecules).

Gel Point

Gelation requires a critical minimum concentration of molecules (C_o) to form a gel network and a critical minimum time (t_c) for the reaction to occur (10). An additional requirement for heat-induced gelation is T_c, the minimal temperature required to cause structural changes which ultimately produce a gel (provided that protein concentration is $\geq C_o$ and reaction time is $\geq t_c$). It should be recalled that it takes one molecule to denature, two or more to aggregate and many more to form a gel matrix. The temperature and time required to form a gel are critical factors in determining if a protein will function in a particular application. For example, the thermal process used to form a meat product may be insufficient for whey protein gelation but adequate for egg albumen gelation.

Factors Determining t_c and T_c. The denaturation temperature and gel point of β-lactoglobulin solutions are altered by NaCl and $CaCl_2$ (Table I). The onset of denaturation (as determined by differential scanning calorimetry) is at 73.4°C and 69.9°C in β-lactoglobulin solutions containing 100 mM NaCl and 20 mM $CaCl_2$, respectively. In addition to the onset of denaturation, gelation occurs at a lower temperature in $CaCl_2$-containing solutions. Gelation occurs at 73.2°C, which is 2°C lower than the maximum transition temperature of denaturation (T_{max}). Addition of $CaCl_2$ to β-lactoglobulin solutions containing 100 mM NaCl caused the gel point temperature to decrease but not to the temperature of $CaCl_2$ alone (Table I). This suggests competing mechanisms between sodium and calcium.

Limited proteolysis will decrease the denaturation temperature and alter the temperature sensitivity of gelation. Hydrolyzing β-lactoglobulin with trypsin, to an extent of 35% of the native protein, destabilizes the protein such that three new denaturation transitions appear at temperatures lower than for the native protein (Table II). Moreover, the gelation rate and maximum gel elastic rigidity (G') are increased at 60°C.

Another way to alter the gelation process is to accomplish various degrees of the fluid phase reactions (Figure 1) at temperatures $> T_c$ then lower the temperature to stop the reactions prior to gelation. This is possible with whey protein solutions when the mineral and protein concentrations are low (6). A 10% (w/v) protein solution of whey protein isolate (WPI) can be heated at 70 - 90°C for up to 80 min without gelation (6). However, the solution will form a gel at 20 - 25°C when $CaCl_2$, $BaCl_2$ or 1,6 hexanediamine is added (6, 11). While the precise mechanism of "pre-heated" protein gelation has not been elicited, it appears that soluble polymers are formed in the initial state and that addition of divalent cations causes formation of a gel network by electrostatic screening, ion bridging or a combination of electrostatic effects with hydrophobic interactions (11, 12). The gel structures are not unique, this is just a separation of the two steps in forming stranded gels (11).

Gelation Reaction

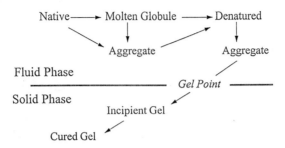

Figure 1. A generalized model for heat-induced gelation of whey proteins based on one proposed in ref. (7).

Table I. Gel Point and Denaturation Temperatures of β-lactoglobulin Solutions Containing Salts[a]

Treatment	Gel Point[b]		$T_o{}^c$ (°C)	$T_{max}{}^d$ (°C)
	Temp., (°C)	Time, (min)		
100 mM NaCl	80	0.68 ± 0.47	73.4 ± 0.16	78.8 ± 0.24
20 mM CaCl$_2$	73.2 ± 0.29		69.9 ± 0.22	75.3 ± 0.05
100 mM NaCl + 20 mM CaCl$_2$	76.6 ± 0.26			
100 mM NaCl + 5 mM CaCl$_2$	77.8 ± 0.23			

[a]Data from ref. (22) for 7% w/v, pH 7.0, β-lactoglobulin solutions.
[b]Gel point represents the temperature of gelation when protein solutions were heated from 25 - 80°C at 1°C/min or the time in subsequent holding at 80°C.
[c]Temperature of denaturation onset from differential scanning calorimeter measurements.
[d]Temperature of maximum transition from differential scanning calorimeter measurements.

Gel Network Types

Globular proteins can be formed into clear, translucent or opaque gels by altering the pH and salt concentration of the protein dispersion (5, 13). The importance of ionic conditions in protein gelation was noted by Ferry (1), who stated that "..the characteristic critical dependence of gelation upon pH, concentration, ionic strength, and the rates of consecutive reactions is evidently common to all these systems of denatured proteins." Bovine serum albumin gels formed at pH > 6.4 and low NaCl concentration are clear (14). Opaque gels are formed at pH values approaching the isoelectric point (pH 5.1), regardless of NaCl concentration. Turbid gels are formed in the intermediate pH range or when salt concentration is increased at pH > 6.4. The diameter of the strands forming the gel network, along with refractive index differences, determines the appearance of globular protein gels. Turbid and opaque gels are formed under conditions where non-specific protein-protein interactions are favored, such as pH close to the pI and high salt concentrations, producing a gel network composed of strands of large protein particles. Bovine serum albumin gels formed at pH 5.1 have matrix strand diameters of 500 nm and greater (5). Clear bovine serum albumin gels formed at pH 6.5 have strand diameters of < 15 nm (5). These observations lead to the definition of gels as *fine-stranded* or *stranded* (those with strand diameters reflecting one or several molecular diameters) or *particulate* (those with strands formed from the aggregation of large particles of protein molecules). *Mixed* gels have networks composed of stranded and particulate structures.

Rheological Properties

Rheological properties are determined at one of two general levels of deformation (strain). Small-strain rheological tests are conducted such that permanent damage to the network is absent or minimal. A non-destructive stress (force/area) or strain (deformation per unit length) is applied and the respective strain or stress is measured. Properties measured are complex modulus (G*, overall rigidity), storage modulus (G', elastic rigidity), loss modulus (G", viscous rigidity) and phase angle (δ, the relative amount of viscous and elastic rigidity). Small-strain tests are used to determine the gel point along with various rheological properties. Large-strain rheological tests deform the material to levels where the matrix is damaged or fractured. If one is interested in factors related to textural properties, then large-strain or fracture rheological properties should be determined. Figure 2 shows a typical stress-strain diagram for torsional fracture of whey protein isolate gels. Properties determined are fracture shear stress (σ_f), fracture shear strain (γ_f) and the fracture modulus (G_f = stress/strain at fracture).

Table II. Denaturation and Gelation of β-lactoglobulin Treated with Limited Proteolysis by Immobilized Trypsin[a]

Treatment	Gelation[b]		$T_o{}^c$	$T_{max}{}^d$
	Time, (min)	G', (Pa)	(°C)	(°C)
Denaturation				
Fragment 1			53.2 ± 1.4	54.0 ± 0.4
Fragment 2			57.7 ± 0.8	58.8 ± 0.6
Fragment 3			63.4 ± 0.5	64.0 ± 0.3
β-lactoglobulin			76.4 ± 0.7	81.8 ± 0.1
Gelation at 60°C				
7% w/v protein				
Intact	755 ± 15	13 ± 0.1		
Hydrolyzed	152 ± 22	176 ± 16		
15% w/v protein				
Intact	342 ± 21	68 ± 15		
Hydrolyzed	116 ± 5	4910 ± 100		

[a]Data from ref. (24).
[b]Gel time is the time required to form a gel and G' values represent the elastic rigidity of the gels after holding for 12 hr at 60°C.
[c]Temperature of denaturation onset from differential scanning calorimeter measurements.
[d]Temperature of maximum transition from differential scanning calorimeter measurements.

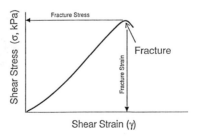

Figure 2. Typical stress vs. strain diagram from torsional fracture of whey protein isolate gels.

Properties Of Gel Types

The physical properties of stranded, particulate and mixed gels are listed in Table III. There are two types of stranded gels which can be formed, both having a translucent appearance and network strands with diameter in the nm range (15). Stranded gels formed at pH > pI (stranded-i) are strong and elastic whereas those formed at pH < pI (stranded-ii) have a weak, brittle structure (16, 17). Mixed gels have maximum strength (fracture stress) and rigidity. Particulate gels have the lowest water-holding properties (Table III).

Effect of Salts on Gel Type and Rheological Properties. The concentration, valence and charge of ions alters the appearance, rheological and water-holding properties of WPI gels by altering the gel type (18, 19). Figures 3-5 show the effects of ion type and concentration on 10% (w/v) protein, pH 7.0, WPI gels. Low concentrations of NaCl (25-30 mM) and $CaCl_2$ (7.5 mM) promote the formation of incipient gels with characteristically low fracture stress and rigidity (Figure 3) (18). Increasing ionic strength from 20 - 100 mM with either NaCl and $CaCl_2$ causes a similar increase in fracture stress; however, fracture rigidity shows cation-specific trends. This is due to the formation of different gel types. Increased concentrations of calcium cause a particulate network to form while increasing the sodium concentration shifts the structure from stranded-i to mixed. The rigidity (stress/strain) of calcium-containing gels decreases because the fracture strain increases (Figure 4).

Increasing sodium phosphate concentration causes rheological transitions similar to an increase in NaCl concentration (Figures 4 and 5). These trends suggest that sodium phosphate and sodium chloride alter strain by a related mechanism(s); in this case the shift from a stranded-i gel network to a mixed network. Fracture rigidity is greatest and fracture strain is minimal with a mixed network structure. In contrast, $CaCl_2$ causes an initial increase in strain as concentration is increased (Figure 4), and fracture rigidity remains constant at concentrations > 30 mM (Figure 5). This effect is also seen with $MgCl_2$ and $BaCl_2$, suggesting a general divalent cation mechanism (18).

The changes in fracture rigidity caused by NaCl, $CaCl_2$ and Na-phosphate can be explained by the transitions from stranded-i to mixed to particulate networks. Gels containing NaCl and Na-phosphate go from zero to peak rigidity as salt concentration increases and the gel networks are transformed from stranded-i to mixed (Figure 5) (19). Mixed gels containing NaCl or Na-phosphate have similar maximum fracture rigidities of 20 - 21 kPa. As salt concentration increases and fracture rigidity decreases, gel networks are transformed from mixed to particulate structures (Figure 5) (19). The parabolic relationship between fracture rigidity and salt concentration means that stranded-i and particulate gels can have the same rigidity (Figure 5). However, these gels can be differentiated based on water-holding (much lower for the particulate gel) and appearance (see Table III). Calcium chloride has a

Table III. Network Types and Physical Properties of Whey Protein Gels[a]

| Gel Type | Fracture Rheology | | | Appearance | Water Holding |
	Stress	Strain	Rigidity		
Stranded					
i) pH > pI	+++[b]	+++	++	Translucent	+++
ii) pH < pI	+	+	++	Translucent	?
Mixed	++++	++	++++	Cloudy/opaque	++
Particulate	+++	+++	++	Opaque	+

[a]Based on refs. (15 - 17, 19).
[b]The number of "+" symbols represents the relative magnitude of a property among gel types, with the greater number of symbols indicating a greater magnitude, and "?" indicated no relevant data.

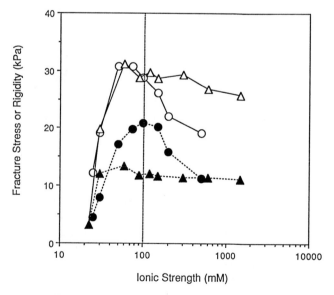

Figure 3. The effect of salt concentration on fracture rigidity (closed symbols) and fracture stress (open symbols) of 10% protein, whey protein isolate gels formed at pH 7.0 from solutions containing NaCl (circles) or $CaCl_2$ (triangles). Data from ref. (18).

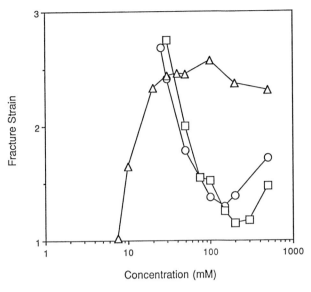

Figure 4. The effect of salt concentration on fracture strain of 10% protein, whey protein isolate gels formed at pH 7.0. Data for sodium (circles) and calcium (triangles) chloride are from ref. (18) and sodium phosphate data (squares) are from ref. (19).

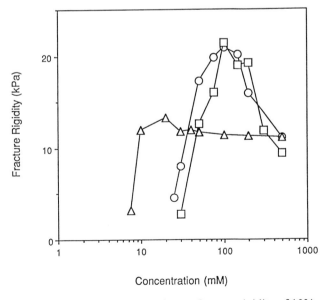

Figure 5. The effect of salt concentration on fracture rigidity of 10% protein, whey protein isolate gels formed at pH 7.0. Data for sodium (circles) and calcium (triangles) chloride are from ref. (18) and sodium phosphate data (squares) are from ref. (19).

different effect, with fracture rigidity reaching a maximum of 12 kPa and remaining relatively constant at higher ionic strengths (Figure 5). Gels containing $CaCl_2$ appear to initially form a weak, mixed structure (note low level of strain in Figure 4) and rapidly progress into a particulate gel. The convergence of fracture rigidities at the highest levels of salts (Figure 5) suggests that a similar, particulate gel is formed, independent of the type of salt used to induce particulate gel formation.

Effects of pH on Gel Type and Rheological Properties. The fracture properties of WPI gels (10% w/v protein) formed from dispersions varying in pH from 2 - 9 are seen in Figure 6. Gels can be separated into three groups based on fracture properties. Those formed at pH 2 and 4 have low fracture stress and strain values and are classified as having a brittle texture. Brittle gels are also formed by β-lactoglobulin in this pH range (16). These gels are classified as having a fine-stranded matrix (15) and, due to the low fracture stress and strain, fit the description for a stranded-ii gel (Table III). The strongest gels, those with the highest fracture stress values, are formed at pH 7 and 8 in dispersions containing 100 mM NaCl. These gels probably have mixed networks. Gels formed at pH 5 and 6, in the presence of sodium or calcium chloride, have rheological properties (medium stress and high strain) which suggest a particulate gel network. This is supported by the observation that β-lactoglobulin forms gels with particulate networks in this pH range (15). The rheological changes associated with increasing the pH from 7-9 in NaCl-containing gels are associated with changing from a mixed (pH 7 & 8) to a stranded-i type gel (note that in going from pH 8 to 9 the stress decreased and the strain increased). From pH 6 to 9 the rheological properties of $CaCl_2$-containing gels were constant and indicative of particulate gels.

Hypothesis

The overall effects of salts and pH on whey protein gelation can be explained based on the gelation reaction model (Figure 1) and gel types (Table III). At low salt concentration and pH > or < pI, denatured proteins form linear aggregates which result in stranded gels. The rheological properties of gels depend on the gel network and intermolecular bonds connecting molecules within the network. Therefore, while stranded-i and stranded-ii gels have a similar appearance, their rheological properties are different due to the amount of intermolecular disulfide bonding (20). Mixed gels appear to be formed by condensing linear strands into larger aggregates, although this mechanism is not established. Particulate gels are formed by several mechanisms. At pH ≈ pI, non-specific protein-protein interactions are favored, resulting in large primary aggregates of native molecules which undergo a second aggregation after denaturation into a gel network. At pH > pI particulate gels are formed at high salt concentrations by extensive condensation of linear aggregates and/or non-specific aggregation of denatured molecules.

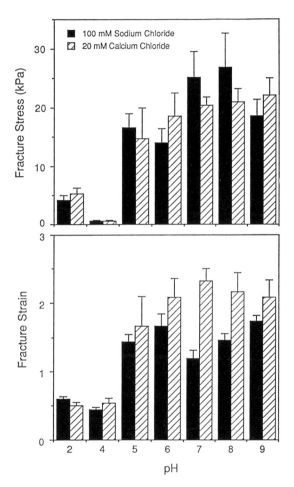

Figure 6. Relationship between pH and stress and strain at fracture for 10% protein, whey protein isolate gels made from solutions containing either 100 mM NaCl or 20 mM CaCl₂. Data from ref. (17).

The effect of $CaCl_2$ depends on whether calcium is present in solution before or after heating. Whey proteins form particulate gels when heated in solutions containing ≥ 20 mM $CaCl_2$ (18, 21). This appears to be caused by an aggregated molten globule state. β-Lactoglobulin solutions at 20-25°C containing $CaCl_2$ or NaCl show no ion-specific changes in structure as measured by circular dichroism (22) or two-dimensional ^{1}H-NMR (12). However, increasing temperature to 40-45°C causes an ion-specific change in two-dimensional ^{1}H-NMR structure (12), and aggregation in β-lactoglobulin and WPI solutions containing $CaCl_2$ (6, 23). This type of aggregation produces a particulate gel when heated to temperature $> T_c$. The type of gel formed from soluble, linear aggregates depends on the rate of mixing. A slow interaction of calcium ions with linear aggregates, as controlled by dialyzing aggregates against $CaCl_2$ solutions (11), forms stranded gels. In contrast, particulate gels can be formed by rapid mixing of $CaCl_2$ with linear aggregates of WPI (unpublished observation). It appears that whey proteins can form a limited set of gel networks by a variety of chemical mechanisms.

Acknowledgments

This work was supported by grants from Dairy Management Inc. and the USDA National Research Initiative Competitive Grants Program.

Literature Cited

1. Ferry, J.D. *Adv. Protein Chem.* **1948**, *4*, 1-78.

2. Xiong, Y.L.; Kinsella, J.E. *J. Agric. Food Chem.* **1990**, *38*, 1887-1891.

3. Færgemand, M.; Otte, J.; Qvist, K.B. *Food Hydrocoll.* **1997**, *11*, 19-25.

4. Otte, J.; Ju, Z.Y.; Færgemand; M. Lomholt, S.B.; Qvist, K.B. *J. Food Sci.* **1996**, *61*, 911-915.

5. Clark, A.H.; Judge, F.J.; Richards, J.B.; J.M. Stubbs; Suggett, A. *Int. J. Peptide Protein Res.* **1981**, *17*, 380-392.

6. Barbut, S.; Foegeding, E.A. *J. Food Sci.* **1993**, *58*, 867-871.

7. Ziegler, G.R.; Foegeding, E.A. *Adv. Food Nutr. Res.* **1990**, *34*, 203-298.

8. Hirose, M. *Trends in Food Sci. Technol.* **1993**, *4*, 48-51.

9. Winter, H.H., Izuka, A. and De Rosa, M.E. *Polymer Gels Networks* **1994**, *2*, 239-245.

10. Ross-Murphy, S.B. In *Food Polymers, Gels and Colloids*, Dickinson, E., Ed.; The Royal Society of Chemistry, Cambridge, UK, 1991; pp. 357-368.

11. Roff, C.F.; Foegeding, E.A. *Food Hydrocoll.* **1996**, *10*, 193-198.

12. Li, H.; Hardin, C.C.; Foegeding, E.A. *J. Agric. Food Chem.* **1994**, *42*, 2411-2420.

13. Doi, E. *Trends in Food Sci. Technol.* **1993**, *4*, 1-5.

14. Richardson, R.K.; Ross-Murphy, S.B. *The British Polymer J.*, **1981**, *13*, 11-16.

15. Langton, M.; Hermansson, A-M. *Food Hydrocoll.* **1992**, *5*, 523-539.

16. Stading, M.; Hermansson, A.M. *Food Hydrocoll.* **1991**, *5*, 339-352.

17. Foegeding, E.A. In *Food Proteins: Structure and Functionality*, Schwenke, K.D.; Mothes, R., Eds.; VCH: Weinheim, Germany, **1993**; pp. 341-343.

18. Kuhn, P.R.; Foegeding, E.A. *J. Agric. Food Chem.* **1991**, *39*, 1013-1026.

19. Bowland, E.L.; Foegeding, E.A. *Food Hydrocoll.* **1995**, *9*, 47-56.

20. Errington, A.D. Rheological and microstructural analysis of whey protein isolate gels: The effect of concentration and pH. M.S. Thesis, North Carolina State University, Raleigh, **1995**.

21. Mulvihill, D.M.; Kinsella, J.E. *J. Food Sci.* **1988**, *53*, 231-236.

22. Foegeding, E.A.; Kuhn, P.R.; Hardin, C.C. *J. Agric. Food Chem.* **1992**, *40*, 2092-2097.

23. Sherwin, C.P; Foegeding, E.A. *Milchwissenschaft*, **1997**, *52*, 93-96.

24. Chen, S.X.; Swaisgood, H.E.; Foegeding, E.A.. *J. Agric. Food Chem.* **1994**, *42*, 234-239.

Chapter 10

Whey Protein Interactions: Effects on Edible Film Properties

John M. Krochta

Department of Food Science and Technology, Department of Biological and Agricultural Engineering, University of California, Davis, CA 95616

Heat denaturing of whey proteins exposes free sulfhydryl groups which promote intermolecular disulfide bond formation, thus allowing formation of insoluble films.   Whey protein isolate (WPI) and ß-lactoglobulin films have similar water vapor permeability (WVP) and oxygen permeability (OP).   Film properties are affected by protein chain-to-chain interactions, which in turn are affected by presence of low-molecular-weight plasticizers.   Plasticizer content and relative humidity (RH) both have an exponential effect on film permeability. At film compositions where glycerol- and sorbitol-plasticized WPI films have equal mechanical properties, sorbitol-plasticized films have lower OP.   Inhibition of intermolecular disulfide bond formation with sodium dodecyl sulfate (SDS) increases WPI film solubility, extendibility and flexibility, while having little effect on WVP. Inhibition of sulfhydryl-disulfide interchange with N-ethyl maleimide reduces WPI film solubility and elongation, with little effect on other film mechanical properties or WVP.   Reduction of disulfide bonds with cysteine has no effect on WVP of WPI films.   Heat curing of glycerol-plasticized WPI films increases film strength, while decreasing film extendibility, flexibility and WVP.   Overall, WPI film properties are affected both by the amount of intermolecular disulfide bond formation and by plasticizers competing for protein chain-to-chain hydrogen bonding.

Edible films based on proteins, polysaccharides and/or lipids have potential for reducing mass transport, reducing mechanical damage, and carrying ingredients such as flavors, colors, antioxidants and antimicrobials in food systems. Films containing lipids can limit migration of moisture when formed or placed between food components which have different water activities.   Protein or polysaccharide films can reduce lipid and flavor migration between food components.   And films formed as coatings on foods can work with packaging to protect food from oxygen and moisture transport, aroma loss and mechanical attrition. Many commercial uses exist and many more are possible (1, 2).

Considerable research in recent years has focused on determination of the properties of films based on proteins (3-5). However, much work remains to understand the structure of protein films and the influence of various structural changes caused by denaturation, moisture, and plasticizers on the properties of films. The objective of this paper is to summarize research in this area for whey protein films.

Formation and Properties of Whey Protein Films

Transparent, flexible, insoluble films can be formed by drying aqueous solutions of heat-denatured whey protein isolate (WPI) and low-molecular-weight plasticizer. Plasticizers are seen as interacting with polymer chains, thus reducing chain-to-chain interaction and thus decreasing film brittleness (i.e., increasing film flexibility and extendibility). In the case of protein and polysaccharide edible films, glycerol, sorbitol and polyethylene glycol have been commonly used as plasticizers (6). These compounds are believed to act mainly by disrupting hydrogen bonding between neighboring polymer chains and thus increasing chain mobility (6, 7).

With the WPI film formulations studied it was necessary to heat the solution (e.g., 90°C for 30 min) to denature the protein (8). Otherwise, films cracked into small pieces when dried, presumably due to the lack of the intermolecular disulfide bonds which are formed during heat denaturation. Optimum films were formed from 8-10% solutions of WPI at pH 7 to 9.

Such WPI films have been found to be excellent oxygen and aroma barriers (9, 10). On the other hand, WPI films are poor moisture barriers, although addition of lipid materials improves their moisture barrier properties (11-13). Table I shows that increasing the relative amount of heat-denatured WPI in films made from emulsions containing beeswax (BW) decreased film water vapor permeability (WVP) (11). With less than 50% heat-denatured WPI, intact films could not be produced. These results appear to indicate that increased intermolecular disulfide bonding reduces WPI film WVP and improves film integrity.

Table I. Denaturing Effect on WVP of WPI Films

Film Composition (WPI:BW:Sor)	WPI Heated*	$WVP \left(\dfrac{g\,mm}{m^2\,kPa\,hr} \right)$	ΔRH
56:00:44	100%	2.14	0/79%
56:28:16	50%	1.72	0/88%
56:28:16	75%	1.18	0/91%
58:28:16	100%	0.86	0/94%

* 90°C, 30 min   WVP = Water Vapor Permeability
SOURCE: Adapted from ref. 11

Effect of Composition and Processing on Film Properties

Comparison of WPI Films with ß-Lactoglobulin Films. The approximate WPI composition, component molecular weights, disulfide bonds and thiol groups are shown in Table II (14). Although ß-lactoglobulin (ß-Lg) makes up more than 50% of WPI, the presence of other components with markedly different molecular weights, number of disulfide bonds and thiol groups, and amino acid sequences in WPI suggests possible different properties for ß-Lg films compared to WPI films.

Table II. Components of Whey Protein

Fraction	MW	S-S	S-H
57% ß-Lg	18,300	2	1
19% α-La	14,200	4	0
7% BSA	66,300	17	1
13% Igs	150,000-1,000,000	*nv	nv
4% PP	41,000-22,000	0	0

*nv = numerous variable
SOURCE: Adapted from ref. 14

However, Figure 1 shows that the water vapor permeabilities (WVP) of WPI and ß-Lg films were not significantly different at each glycerol plasticizer content studied (15). On the other hand, glycerol content had a significant effect on WVP for both WPI and ß-Lg. Although there appears to be a nearly linear relationship between glycerol content and WVP, films with the lowest glycerol content experienced the highest relative humidity (RH) in the test (84%) while films with the highest glycerol content experienced the lowest RH (68%) in the test. This is an important issue, as the WVP of whey protein films increases with film moisture content (See below). Thus, if it were possible to design the test to expose all the films to the same RH, an exponential-type relationship between WVP and glycerol content would likely be experienced for both WPI and ß-Lg films.

Figure 2 shows the exponential relationship found between oxygen permeability (OP) and glycerol content at each temperature studied with constant 40% RH (15). There was no significant difference between WPI and ß-Lg films. Temperature was also found to have an exponential effect on OP.

Thus, in spite of the different film molecular structures in WPI films compared to ß-Lg films and the relative differences in hydrogen bonds, hydrophobic interactions and disulfide bonds, no differences in WVP and OP were found. The amount of glycerol was certainly more important than the whey protein component.

Table III. Relative Humidity Effect on WVP of WPI Films

Relative Humidity	WPI:Sorbitol = 1.00 WVP ($\frac{g\ mm}{m^2\ kPa\ hr}$)	WPI:Glycerol = 1.67 WVP ($\frac{g\ mm}{m^2\ kPa\ hr}$)
10%	0.2	0.2
32%	0.6	0.4
40%	-----	0.6
50%	0.9	1.6
60%	-----	3.9
65%	2.3	5.0
75%	3.5	-----

SOURCE: Adapted from ref. 8   WVP = Water Vapor Permeability

Effect of Relative Humidity. As mentioned above, moisture also acts as a plasticizer for whey protein films. The results of Table III suggest an exponential-

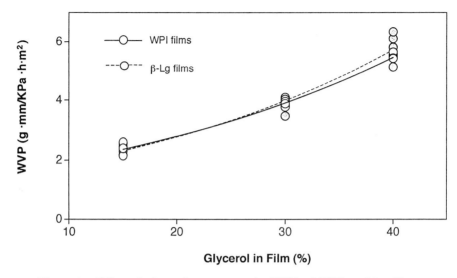

Figure 1. Effect of glycerol content on the WVP of WPI or ß-Lg films. (Reproduced with permission from ref. 15. Copyright 1996 American Chemical Society.)

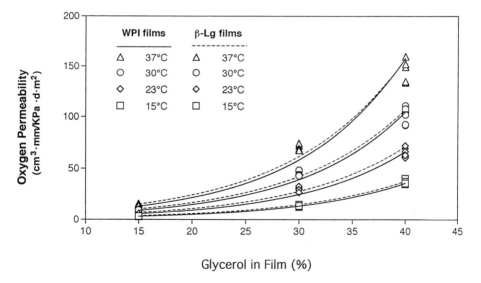

Figure 2. Effect of glycerol content on the OP of WPI or ß-Lg films. (Reproduced with permission from ref. 15. Copyright 1996 American Chemical Society.)

type relationship between WVP and RH for WPI films (8). Interestingly, the effect of RH on WVP is more pronounced for glycerol-plasticized films compared to sorbitol-plasticized films, in spite of higher WPI:glycerol ratio compared to WPI:sorbitol ratio. Similar effect of RH on OP of WPI films has been observed (9).

These results indicate the importance of considering RH when comparing properties of different edible films. They also show the importance of RH when considering possible applications of edible films to food products.

Effect of Plasticizer Type. Ideally, a plasticizer is available for each film type which decreases film brittleness while not markedly increasing film permeability. Plasticizers studied include polyols (e.g., glycerol, sorbitol and polyethylene glycol) and surfactants (e.g., fatty acids, sodium dodecyl sulfate, monoglycerides).

Polyols. The permeabilities of films with different plasticizers are often compared to assess plasticizer effect on permeability. Such a comparison is made in Table IV, which shows that at equal concentration sorbitol has less effect on increasing WPI film WVP than does glycerol (8). Such comparisons are useful but not as helpful as when simultaneously assessing impact of plasticizer type on film mechanical properties. An integrated approach involving determination of film permeability and mechanical properties in parallel is necessary for complete assessment of plasticizer type and amount.

Table IV. Plasticizer Effect on WVP of WPI Films

Plasticizer[a]	WVP ($\frac{g\ mm}{m^2\ kPa\ hr}$)	ΔRH
50% Glycerol	6.4	0/59%
38% Glycerol	5.0	0/65%
50% Sorbitol	3.5	0/75%
38% Sorbitol	2.6	0/79%

[a] % dry basis   WVP = Water Vapor Permeability
SOURCE: Adapted from ref. 8

Table V compares the OP and mechanical properties of WPI films containing different amounts of glycerol and sorbitol (9). Films containing 30% glycerol had similar tensile strength (TS) to films containing 30 and 50% sorbitol. However, the sorbitol-plasticized films had 93% and 89% lower OP, respectively. Films containing 15% glycerol had similar elongation (E), a measure of film extendibility, to films containing 30, 40 or 50% sorbitol. However, once again the sorbitol-plasticized films had much lower OP than the comparable glycerol-plasticized films, ranging 56-72% lower. Finally, films containing 15 and 30% glycerol had elastic modulus (EM), similar to films containing 30 and 50% sorbitol, respectively. The EM is a measure of film stiffness or lack of film flexibility. The comparable sorbitol-plasticized films had 70 and 90% lower OP, respectively.

These results suggest that at the same mass content levels in films, lower molecular weight plasticizers may result in films with higher permeability, flexibility and extendibility. This is perhaps understandable, considering that the molar content of the lower molecular weight plasticizer in the film is greater in such a comparison. However, lower molecular weight plasticizers also appear to increase permeability more when present in films at levels giving comparable strength, flexibility and extendibility to higher molecular weight plasticizers. More research is needed to understand the effects of plasticizers on films so that film properties can be optimized as to permeability and mechanical properties.

Table V. Plasticizer Effect on Properties of WPI Films

Plasticizer[a]	$OP \left(\frac{cc\ \mu m}{m^2\ kPa\ hr}\right)$	TS (MPa)	E (%)	EM (MPa)
15% Glycerol	18	29	4	1100
30% Glycerol	76	14	31	475
30% Sorbitol	5	14	3	1040
40% Sorbitol	6	18	5	625
50% Sorbitol	8	15	8	475

[a] % dry basis   TS = Tensile Strength
OP = Oxygen Permeability   E = Elongation (Extendibility)
SOURCE: Adapted from ref. 9   EM = Elastic Modulus (Stiffness)

Sodium Dodecyl Sulfate. Surfactants are used less frequently as plasticizers in edible films, because they have been found less effective than polyols (16). However, the ability of sodium dodecyl sulfate (SDS) to associate with and denature proteins based on hydrogen bond and hydrophobic interaction disruption and charge repulsion suggested its investigation as a plasticizer for WPI films (17). SDS was added to the protein solution before heating to possibly reduce the amount of intermolecular interaction occurring due to protein unfolding.

Although SDS could not plasticize WPI films by itself, it did produce interesting results when used in combination with sorbitol or glycerol. Table VI shows the effect of increasing levels of SDS in films with WPI:sorbitol of 3:1. Up to SDS:WPI of approximately 0.01 (g/g), there was little effect on the film permeability, flexibility and solubility. However, at SDS:WPI of 0.2, there was significantly lower film strength and greater film extendibility, flexibility and solubility, with no effect on permeability. Thus, SDS helped achieve favorable mechanical properties (specifically, improved extendibility and flexibility) with no increase in permeability. In this sense, it was a quite effective co-plasticizer. The increase in film solubility at SDS:WPI>0.01 suggests that the effect of SDS is due primarily to reducing the amount of intermolecular disulfide bond formation.

Table VI. SDS Effect on Properties of WPI:Sorbitol (3:1) Films

SDS:WPI (g/g)	$WVP \left(\frac{g\ mm}{m^2\ kPa\ hr}\right)$	TS (MPa)	E (%)	EM (MPa)	S (%)
0.0001	3.0	5.5	30	500	25
0.001	2.8	5.5	35	500	26
0.01	2.9	4.4	14	460	29
0.1	3.2	4.0	53	340	38
0.2	3.0	3.3	65	240	59

WVP = Water Vapor Permeability   E = Elongation (Extendibility)
TS = Tensile Strength   EM = Elastic Modulus (Stiffness)
SOURCE: Adapted from ref. 17   S = Solubility

The effect of SDS:WPI of 0.2 was similar in films with WPI:glycerol of 3:1 (Table VII). Films had significantly better extendibility and flexibility with no significant increase in permeability. However, at SDS:WPI greater than ~0.2 there was a reverse in film extendibility and flexibility with a decrease in permeability. This suggests that at these higher levels, the SDS effect was more due to its being a large fraction of the film composition relative to its effect on WPI.

Table VII. SDS Effect on Properties of WPI:Glycerol (3:1) Films

SDS:WPI (g/g)	WVP ($\frac{g\,mm}{m^2\,kPa\,hr}$)	TS (MPa)	E (%)	EM (MPa)	S (%)
0	3.6	9.2	14	400	12
0.2	3.9	3.5	46	110	27
0.4	3.7	5.2	21	170	50
0.6	3.1	4.9	13	190	54
0.8	2.9	4.3	8	205	65
1.0	2.6	4.0	5	205	58

WVP = Water Vapor Permeability   E = Elongation (Extendibility)
TS = Tensile Strength   EM = Elastic Modulus (Stiffness)
SOURCE: Adapted from ref. 17   S = Solubility

Overall, the results with SDS appear to suggest that film mechanical properties are affected both by limiting the amount of intermolecular disulfide bond formation allowed to occur upon denaturation and by adding plasticizers to compete for intermolecular chain-to-chain hydrogen bonding after denaturing is complete. Controlling the amount of intermolecular disulfide bond formation may be possible by adjusting the time and temperature of heat denaturing and/or the use of SDS. Of important consideration is the degree of film insolubility desired in possible food applications.

N-**Ethylmaleimide.** Considering the interesting results obtained with SDS, WPI films were prepared by adding N-Ethylmaleimide (NEM) to neutral pH solutions of WPI before heat denaturing at 90°C for 30 min (18). NEM derivatizes free thiol groups and thus prevents sulfhydryl/disulfide interchange. Such blocking would presumably reduce the extent of intermolecular disulfide bond formation in resulting films. Varying amounts of NEM were added to block from 0 to 100% (theoretical calculation) of free thiol groups. Such films would not be edible, but greater understanding of the role of sulfhydryl/disulfide bond interchange in edible films was the goal. After heat treatment, sufficient glycerol was added to produce films with WPI:Gly of 3:1.

Surprisingly, little effect on film properties occurred over the range of NEM used (Table VIII). WVP did not change; and rather than increasing film extendibility, flexibility and solubility, flexibility did not change significantly, and extendibility and solubility actually decreased. These results indicated that inhibition of sulfydryl/disulfide bond interchange likely produced greater hydrogen bonding or hydrophobicity in the resulting films, with resulting intermolecular interactions giving less film extendibility and solubility.

Table VIII. NEM Effect on Properties of WPI:Glycerol (3:1) Films

SDS:WPI (g/g)	WVP ($\frac{g\ mm}{m^2\ kPa\ hr}$)	TS (MPa)	E (%)	EM (MPa)	S (%)
0	3.7	6.5	20	340	21
0.01	3.6	8.0	23	390	19
0.02	3.5	7.5	20	415	10
0.03	3.5	6.5	15	370	12
0.04	3.5	6.5	11	350	11

WVP = Water Vapor Permeability   E = Elongation (Extendibility)
TS = Tensile Strength   EM = Elastic Modulus (Stiffness)
SOURCE: Adapted from ref. 18   S = Solubility

Cysteine. WPI films were also prepared with addition of cysteine to WPI solutions, both before and after heat denaturing at 90°C for 30 min (*18*). Cysteine reduces disulfide bonds and was expected to prevent or reduce intermolecular disulfide bonds. Varying amounts of cysteine were added from 0 to 150% (theoretical calculation) of that necessary to reduce disulfide bonds in the starting WPI.

Table IX. Cysteine Effect on WVP of WPI:Glycerol (3:1) Films

Cysteine:WPI (mmol/g)	WVP ($\frac{g\ mm}{m^2\ kPa\ hr}$)	S-S Reduced (theoretical)
0	3.7	0%
0.2	3.8	62%
0.3	3.6	93%
0.4	3.6	124%
0.5	3.5	155%

SOURCE: Adapted from ref. 18   WVP = Water Vapor Permeability

Cysteine had no effect on WVP of resulting films, whether added before or after heat denaturing of WPI (Table IX). Since cysteine can react with disulfide bonds in various ways, this result can be interpreted in different ways. However, one interpretation is that WPI intermolecular disulfide bonds do not seem to influence WPI film WVP.

Effect of Film Heat-Curing. Heat curing or irradiation is often used to crosslink synthetic polymers, with resulting improvement in polymer strength and barrier properties. Heat curing of WPI films was seen as an approach to study the effect of increased intermolecular disulfide bonding on film properties (*19*). Films were made from solutions of WPI which had been heated at 90°C for 30 min before adding glycerol so that WPI:glycerol was 3:1. Films dried at room conditions were then heat cured at several temperatures and RHs.

Table X shows the effect of heat curing at 80°C and 60% RH for different times. Films were conditioned to the same temperature and RH before testing. Clearly, increased crosslinking produced films with greater strength and lower extendibility, flexibility and WVP. Similar results were obtained at different temperatures and RHs, with larger temperatures and lower RHs generally having greatest effect.

Table X. Heat Curing Effect on Properties of WPI:Glycerol (3:1) Films

Cure Time (hr)	WVP ($\frac{g\ mm}{m^2\ kPa\ hr}$)	TS (MPa)	E (%)	EM (MPa)
0	3.7	11	33	370
2	3.3	20	19	700
6	3.2	26	9	950
24	2.1	36	5	1240
48	1.7	45	3	1750

WVP = Water Vapor Permeability   TS = Tensile Strength
SOURCE: Adapted from ref. 19   E = Elongation (Extendibility)
                                     EM = Elastic Modulus (Stiffness)

Similar to the results with heat denaturing of WPI solutions and addition of SDS to WPI solutions, these results appear to suggest that film properties are affected both by the amount of intermolecular disulfide bond formation and by plasticizers competing for intermolecular chain-to-chain hydrogen bonding. In addition to heat denaturing of WPI solutions and addition of SDS, heat curing provides another possible approach to controlling the amount of intermolecular disulfide bond formation in films.

Conclusions

Whey protein films are excellent oxygen and aroma barriers and poor moisture barriers. Due to their ability to increase polymer chain mobility, plasticizers and RH have a large effect on film properties, generally increasing film extendibility, flexibility and permeability while decreasing film strength. Film properties are quite dependent on plasticizer type and amount. WPI film properties are also quite dependent on extent of intermolecular disulfide bond crosslinking, which decreases film extendibility, flexibility, permeability and solubility, while increasing film strength. Additional research is needed to determine the combination of WPI intermolecular disulfide bonding and plasticizer type and amount to optimize film properties.

Literature Cited

1. Krochta, J. M. In *The Wiley Encyclopedia of Packaging Technology, 2nd Edition;* Brody, A. L.; Marsh, K. S., Eds.; John Wiley & Sons, Inc.: New York, 1997; pp 397-401.
2. Krochta, J. M.; De Mulder-Johnston, C. L. C. *Food Technology* **1997,** *51*(2), 61-74.
3. Gennadios, A.; McHugh, T. H.; Weller, C. L.; Krochta, J. M. In *Edible Coatings and Films to Improve Food Quality;* Krochta, J. M.; Baldwin, E. A.; Nisperos-Carriedo, M., Eds.; Technomic Publishing Co., Inc.: Lancaster, PA, 1994; pp 201-277.
4. Torres, J. A. In *Protein Functionality in Food Systems;* Hettiarachchy, N. S.; Zeigler, G. R., Eds.; Marcel Dekker, Inc.: New York, 1994; pp 467-507.
5. Krochta, J. M. In *Food Proteins and their Applications in Foods;* Damodaran, S.; Paaraf, A., Eds.; Marcel Dekker, Inc.: New York, 1997; pp 529-549.
6. Guilbert, S. In *Food Packaging and Preservation: Theory and Practice;* Mathlouthi, M., Eds.; Elsevier Applied Science Publishers: New York, 1986; pp 371-94.

7. Kester, J. J.; Fennema, O. R. *Food Tech.* **1986,** *40*(12), 47-59.
8. McHugh, T. H.; Aujard, J. F.; Krochta, J. M. *J. Food Sci.* **1994,** *59*(2), 416-419, 423.
9. McHugh, T. H.; Krochta, J. M. *J. Agric. Food Chem.* **1994,** *42*(4), 841-845.
10. Miller, K. S.; Krochta, J. M. *J. Food Sci.* **In press.**
11. McHugh, T. H.; Krochta, J. M. *J. Am. Oil Chem. Soc.* **1994,** *71*(3), 307-12.
12. McHugh, T. H.; Krochta, J. M. *J. Food Proc. Preserv.* **1994,** *18*(3), 173-88.
13. Shellhammer, T. H.; Krochta, J. M. *J. Food Sci.* **1997,** *62*(2), 390-394.
14. Dybing, S. T.; Smith, D. E. *Cult. Dairy Prod. J.* **1991,** *57*, 4-12.
15. Maté, J. I.; Krochta, J. M. *J. Agric. Food Chem.* **1996,** *44*(10), 3001-3004.
16. Gontard, N.; Duchez, C.; Cuq, J.-L.; Guilbert, S. *Intl. J. Food Sci. Tech.* **1994,** *29*, 39-50.
17. Fairley, P.; Monahan, F. J.; German, J. B.; Krochta, J. M. *J. Agric. Food Chem.* **1996,** *44*(2), 438-443.
18. Fairley, P.; Monahan, F. J.; German, J. B.; Krochta, J. M. *J. Agric. Food Chem.* **1996,** *44*(12), 3789-3792.
19. Miller, K. S.; Chiang, M. T.; Krochta, J. M. *J. Food Sci.* **In press.**

Chapter 11

Thermal Denaturation and Gelation Characteristics of β-Lactoglobulin Genetic Variants

Joyce I. Boye[1], Ching - Y. Ma[2], and Ashraf A. Ismail[3]

[1]Food Research and Development Centre, Agriculture and Agri-Food Canada, 3600 Casavant Boulevard West, St. Hyacinthe, Quebec J2S 8E3, Canada
[2]Department of Botany, University of Hong Kong, Pokfulam Road, Hong Kong
[3]Department of Food Science and Agricultural Chemistry, Macdonald Campus of McGill University, Ste-Anne-de-Bellevue, Montreal, Quebec H9X 3V9, Canada

The thermal characteristics of β-lactoglobulin (β-lg) genetic variants A and B were studied at pH 3.0, 5.0, 7.0 and 8.6 by differential scanning calorimetry (DSC). The β-lg B exhibited higher denaturation temperature and enthalpy than β-lg A and also denatured in a more cooperative fashion as indicated by a lower width at half-peak height. Fourier transform infrared spectroscopy (FTIR) was used to monitor changes in secondary structure of the two proteins when heated from 25 to 95 °C. Results showed that β-lg A had a lower β-sheet content than the B variant at pH 3.0 and 5.0. At pH 7.0 and 8.6 the secondary structure of the two variants were similar. Aggregation bands (1682 cm^{-1} and ~1622 cm^{-1}) were observed when the proteins were heated at all pH values. The microstructure of gels made from 10% (w/v) solutions of β-lg A and B heated at 90 °C for 30 min was studied by electron microscopy. The gel matrix of β-lg B at both acidic and alkaline pH was found to be made up of larger aggregates than the A variant. The aggregates of both variants were large (1-2 μm) and globular at acidic pH but much smaller (nanometer range) and amorphous at alkaline pH.

Heat-induced gelation of whey proteins is an important functional characteristic in the development of a variety of food products. Gel formation of food proteins has been viewed as a two-stage sequential process involving initial denaturation followed by subsequent protein-protein and crosslinking interactions that result in the formation of a three-dimensional gel matrix (1). Several factors influence the gelling properties of food proteins. These include their amino acid composition and sequence, their secondary and tertiary conformation in solution as well as their stability to thermal denaturation. β-lg is the major gelling protein in whey. Seven genetic variants of this protein have been identified which differ in their amino acid composition and sequence. The most prevalent of these variants in commercial preparations are the A and B which vary in positions 64 and 118 where the aspartic acid and valine of β-lg A are replaced by glycine and alanine in β-lg B. Replacement of one amino acid with another can have profound effects on protein functionality, particularly, when the amino acids are from different groups as in β-lg A and B (i.e., the aspartic acid in β-lg A is a dicarboxylic acid while glycine in β-lg B is an aliphatic amino acid). There is a growing need in the food industry for ingredients with specific functional characteristics; this makes it necessary to investigate the peculiar properties of protein genetic variants for their application in specific foods. Research done to date has indicated that the seemingly minor difference in the amino acid composition of β-lg A and B has significant effects on their thermal stability, structural flexibility and susceptibility to proteolysis (2,3). In this study we determine the differences in the thermal properties and secondary structure of β-lg A and B as a function of pH as well as differences in the microstructure of their gels formed at acid and alkaline pH.

Materials and Methods

Materials. β-Lactoglobulin A (L-7880) and B (L-8005) were purchased from Sigma Chemical Co. (St. Louis, Mo) and used as received. Deuterium oxide (D_2O) (product 15, 188-2) was purchased from Aldrich (Milwaukee, WI). All other reagents were of analytical grade.

Sample Preparation. Solutions of β-lg A and B (10% w/v) were prepared by dispersing the proteins in phosphate buffers at pH 3.0, 5.0, 7.0 and 8.6 (ionic strength 0.2ID). For the FTIR study, the phosphate buffers were prepared using D_2O instead of H_2O because D_2O has greater transparency in the region of interest (1600-1700 cm^{-1}). Reagents used for preparation of phosphate buffers were H_3PO_4, KH_2PO_4 and Na_2HPO_4. To form gels, aliquots (0.5 mL) of the β-lg A and B solutions, were placed in microcentrifuge tubes and heated for 30 min at 90°C. The gels formed were allowed to equilibrate at 4°C for 24 h and their microstructures were examined by transmission electron microscopy.

Fourier Transform Infrared Spectroscopy. A Nicolet 8210 FTIR spectrometer (Nicolet Instrument Corporation, Madison, WI), equipped with a deuterated triglycine sulfate detector, was used to record infrared (IR) spectra of the β-lg solutions. The samples were held in an IR cell with a 25 μm pathlength and CaF_2 windows. The temperature of the sample was regulated by placing the cell in a thermostated holder and employing an Omega temperature controller (Omega Engineering, Laval, QC, Canada). The temperature was increased in 5°C increments and the cell allowed to equilibrate for 3 min prior to data acquisition. The reported temperatures are accurate within ± 0.5°C. Five hundred and twelve scans were averaged at 4 cm^{-1} resolution. Wavenumber accuracy was within ± 0.01 cm^{-1}. The spectrometer was purged with dry air from a Balston dryer (Balston, Haverhill, MA). Deconvolution (4) of the observed spectra were performed using the Nicolet FTIR Software, Omnic 1.2a. The signal to noise ratio was >20,000:1, and the bandwidth used for deconvolution was 13 cm^{-1} with a narrowing factor of 2.4. All FTIR experiments were done in duplicates.

Differential Scanning Calorimetry. A DuPont 1090 thermal analyzer (DuPont Co., Wilmington, DE) equipped with a 910 DSC cell base and a high pressure cell was used to measure the thermal properties of the β-lg A and B solutions prepared at different pH values. To evaluate the effect of protein concentration on thermal stabilities of the two variants a DSC 2910 modulated DSC from TA Instruments (Newcastle, DE) was used. (A different DSC was used for the concentration studies because of availability. Results from both instruments were, however, compared for accuracy and reproducibility). Aliquots (10 μl) of each solution were placed in preweighed DSC pans, hermetically sealed and weighed accurately. A sealed empty pan was used as a reference. The samples were heated from 20°C to 120°C at a programmed heating rate of 5°C/min. Indium standards were used for temperature and energy calibrations. Starting temperature of the DSC peak (Ts), peak temperature (T_D), width of peak at half-height (T_w) and heat of transition or enthalpy (ΔH), were computed. (ΔH values were based on actual protein content of the solution placed in the DSC pan). All DSC measurements were made in triplicates.

Electron Microscopy. The β-lg A and B gels were cut into 0.5 mm cubes, fixed in 2% aqueous glutaraldehyde for 24 h at 6°C, and post fixed in 2% osmium tetroxide in 0.05M veronal-acetate buffer and 0.2M imidazole (1:1 ratio of buffer to imidazole), adjusted to pH 6.75. Imidazole was used to prevent the development of 'pepper artifact' by the incidence of minute black dots in the micrograph to which milk proteins are susceptible (5). The samples were then washed with veronal-acetate buffer and dehydrated in a graded ethanol series (20%, 40%, 60%, 80%, 95% and absolute ethanol). The dehydrated samples were embedded in medium hard SPURR's low-viscosity medium and sectioned. Sections (90 nm thick), obtained using a diamond knife mounted on a Leica Ultracut microtome, were stained with uranyl acetate and lead citrate solutions. Each sample was embedded in duplicate and six areas were examined in each block. Micrographs were taken on 35-mm film in a Zeiss EM 902 transmission electron microscope operated at 80 kV.

Results and Discussion

FTIR. Figure 1a shows the deconvoluted infrared spectra of β-lg A and B at pH 3.0. The spectra of both proteins showed seven principal bands at 1692 (β-type structure), 1681 (β-sheet/turns), 1669 (turns), 1649 (α-helix), 1636/7 (β-sheet), 1629/28 (β-strand) and 1614 cm^{-1} (side-chain vibrations), (6-9). Examination of the spectra showed the two proteins to have different secondary structures at pH 3.0. The characteristic difference between the two spectra at pH 3.0 was in the intensities of the bands at 1681, 1649 and 1637 cm^{-1}. In the spectrum of β-lg B the intensities of these bands were much higher suggesting that the B variant had a greater amount of α-helical and β-sheet structure than the A variant. The intensity of the 1629 cm^{-1} band was, however, lower for the B variant suggesting that the A had a great amount of β-strand content. As pH was increased to 8.6 (Figure 1b) a marked increase in β-sheet formation (1635 cm^{-1} band) was observed for both variants, accompanied by a relative decrease in β-strand content (1625/24 cm^{-1}). The 1681, 1669, 1649, 1637 and 1629 cm^{-1} bands in the spectra of both variants also shifted by 2-5 cm^{-1} to lower wavenumbers which is

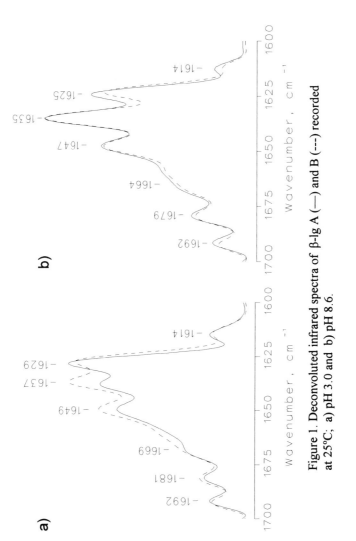

Figure 1. Deconvoluted infrared spectra of β-lg A (—) and B (---) recorded at 25°C; a) pH 3.0 and b) pH 8.6.

indicative of an increase in hydrogen bonding (7). At pH 8.6 (Figure 1b) and 7.0 (spectra at pH 7.0 not shown) both variants appeared to have very similar secondary structures. Although the A and B forms differ at only two positions, the substitution of a glycine for an aspartic acid residue at position 64 in the A form is significant because of the increased likelihood of an extra salt bridge forming between the carboxylic group of the Asp residue and any other basic group in the monomer unit. At pH 3.0, the carboxyl group of the Asp residue is not ionized; this would tend to decrease self-association and may explain the decrease in the β-sheet content of the A variant at pH 3.0. At alkaline pH's, ionization of the carboxyl group of the Asp residue would tend to increase self-association which might explain the increase in β-sheet content observed at alkaline pH for both variants.

Figure 2a shows the changes that occurred in the secondary structure of β-lg A when heated from 25°C to 70°C at a pH close to its isoelectric point (pH 5.0) (similar results were also observed at pH 3.0). The major changes observed during heating were a decrease in the 1692 (indicative of unfolding), 1649 (loss of α-helical structure), 1629 (loss of β-strand) and 1614 cm^{-1} bands. A general decrease in the intensity of the bands in the 1600 - 1500 cm^{-1} (Amide II) region was also observed; a decrease in this region is indicative of hydrogen-deuterium exchange as the protein unfolds and allows D$_2$O to exchange with hydrogen in the interior of the molecule (i.e., N-H to N-D transition). In contrast, there was an increase in the intensity of the 1676 cm^{-1} band (turns) and an initial increase in the 1636 cm^{-1} band (β-sheet) which decreased consistently when the protein was heated above 60°C. Similar results were observed for the B variant except that there was no initial increase in β-sheet content. Thermal denaturation of β-lg follows a series of unfolding and association steps. The results obtained suggest three stages in the unfolding and subsequent aggregation of β-lg A and B at acid pH. The initial phase (between 25 and 60°C) involves an increase in β-sheet formation (this was absent for the B variant) accompanied by a concomitant decrease in β-strand content, possibly due to association of the protein molecules. Further heating results in partial dissociation of the β-sheet structure possibly into β-strands. In the third phase (above 75°C) (Figure 2b), there is further loss of both β-sheet and β-strand structures as new intermolecular anti-parallel β-sheet structures, (bands at 1682 and 1622 cm^{-1}) are formed.

A comparison of the spectra of β-lg A and B heated from 25 to 85°C at pH 3.0 and 8.6 is shown in Figure 3. At pH 3.0, heating both β-lg A and B to 80°C resulted in the complete disappearance of the 1692 cm^{-1} band (Figure 3a, b). The disappearance of this band in the infrared spectra of β-lg AB has been previously associated with the onset of unfolding (9). For the A variant, the disappearance of the 1692 cm^{-1} band at 80°C was accompanied immediately by the total breakdown in secondary structure (marked decrease in the intensities of the 1681, 1649, 1637 and 1629 cm^{-1} bands) indicative of complete denaturation (9). For the B variant, however, the loss of secondary structure did not occur until the temperature was raised to 85°C, suggesting that it is thermally more stable and required a higher temperature to completely denature. The loss of the major bands in the spectra of both β-lg variants resulted in the appearance of the bands at 1682 and 1622 cm^{-1}. These two bands have been attributed to the formation of intermolecular hydrogen-bonded anti-parallel β-sheet structures associated with aggregate formation (9, 10).

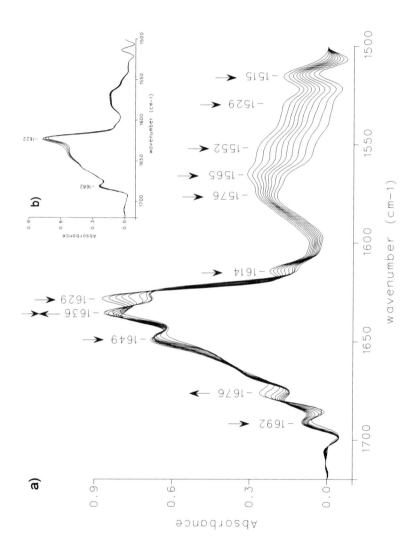

Figure 2. Overlapping infrared spectra of β-lg A at pH 5.0 heated from (a) 25 to 70°C, (b) 80 to 95°C.

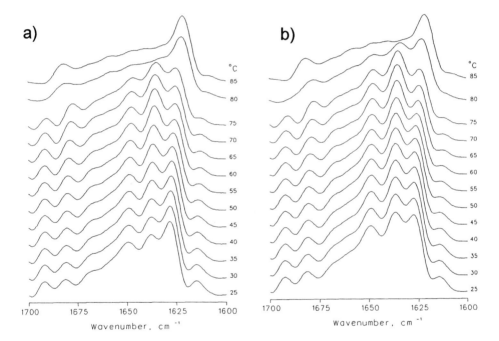

a)

°C
85
80
75
70
65
60
55
50
45
40
35
30
25

Wavenumber, cm⁻¹

1700   1675   1650   1625   1600

b)

°C
85
80
75
70
65
60
55
50
45
40
35
30
25

Wavenumber, cm⁻¹

1700   1675   1650   1625   1600

Figure 3. Stacked plot of deconvoluted infrared spectra of β-lg A and B in deuterated phosphate buffer at pH 3.0 (a,b) and 8.6 (c,d). a,c: β-lg A; b,d: β-lg B. (Adapted from ref. 20).

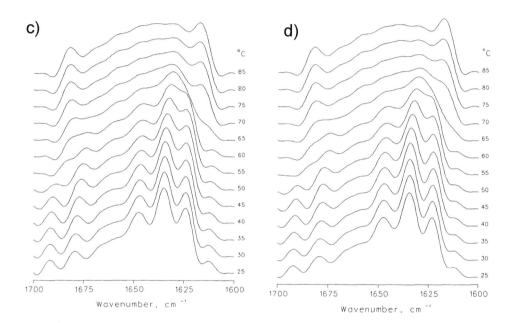

Figure 3. *Continued.*

At pH 8.6 (Figure 3c,d), the 1692 cm^{-1} band in the spectra of both genetic variants completely disappeared at 55°C. This temperature represents a decrease of about 25°C from that observed at pH 3.0. The results indicate that both β-lg A and B are thermally less stable at alkaline pH. Between 55 and 65°C, at pH 8.6, the bands at 1647, 1635 and 1625/24 cm^{-1} in the spectra of both variants become increasingly broad suggesting a change in their secondary structures. The two bands associated with aggregate formation were observed at 1682 and 1618 cm^{-1} when the proteins were heated above 70°C. The results indicate that between 55 and 70°C the proteins assume a configuration which is different from that of the native state and the aggregated state. This intermediary state may represent the molten globule stated described by McSwiney et al. (*11*). It is also evident from the spectra at 70°C (Figure 3c and d, pH 8.6) that β-lg A aggregated faster than β-lg B (i.e., faster rate of increase of 1618 cm-1 band). Formation of aggregate structures upon heat treatment of β-lg has been generally attributed to hydrophobic and thiol-disulfide exchange reactions (*12*).

Differences in the rate of aggregation of β-lg A and B at acid and alkaline pH can be seen in Figure 4. Figure 4a shows the rate of increase in the intensity of the aggregation band (1616 - 1622 cm^{-1}) of β-lg A at pH 3.0, 5.0, 7.0 and 8.6 as a function of temperature. The graph clearly shows that aggregation occurs much faster at pH 7.0 and 8.6 than at pH 3.0 and 5.0. A similar trend was observed for β-lg B. Figure 4b shows a comparison of the rate of aggregation of β-lg A and B at pH 3.0 and 8.6. The graph shows the rate of aggregation of β-lg A at pH 8.6 to be faster than β-lg B. At pH 3.0, the A variant still aggregated faster than the B between 75 to 83°C. Above this temperature, the B variant appeared to form more or larger aggregates.

DSC. The effect of pH on the DSC characteristics (T_s, T_D, T_w and ΔH) of β-lg A and B is presented in Table I. In general, the T_s and T_D values at all four pH's were higher for variant B than variant A, suggesting that the B variant was thermally more stable. This confirms the findings of the FTIR which also showed the B variant to be more thermally stable. The width of the peak at half-height (T_w) of variant A was generally greater than that for variant B at both acid and alkaline pH. In addition, T_w for both proteins was greater at pH 8.6 than at pH 3.0. T_w has been used as an indication of the cooperativity of protein unfolding (*13*). Progressive sharpening of the endothermic peak (decrease in T_w) is indicative of an increase in cooperativity. This suggests that denaturation of β-lg A is less cooperative than β-lg B and also that the denaturation of both β-lg variants was more cooperative at acid pH than at alkaline pH. The lowest T_w for both variants was observed at pH 5.0. This pH is close to the isoelectric point of β-lg. At the isolectric point, protein molecules have the greatest number of attractive forces and normally precipitate out of solution. β-lg is one of the few globular proteins that remains in solution near its isoelectric point (*14*). The molecular state of the proteins at this pH may explain the greater degree of cooperativity observed. The enthalpy values observed for variant B at all four pH's were also greater than those for variant A which suggests that β-lg B, in addition to being more thermally stable, required a greater amount of energy to denature. Partial unfolding of protein molecules generally results in a lowering of ΔH, since the partially denatured proteins require less heat to denature than native proteins. The results therefore suggest that the proteins may have been partially denatured at pH 8.6 prior to heat treatment, and both variants were denatured to about the same extent.

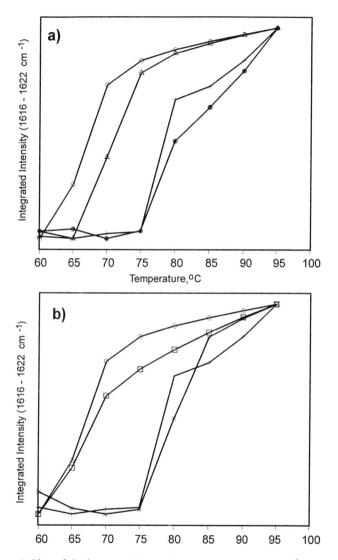

Figure 4. Plot of the integrated intensity of the 1616-1622 cm^{-1} band in the infrared spectra of a) β-lg A at pH 3.0 (+), 5.0 (●), 7.0 (Δ) and 8.6 (○); b) β-lg A, pH 3.0 (+); β-lg B, pH 3.0 (✳); β-lg A, pH 8.6 (○); β-lg B, pH 8.6 (□).

There has been contradicting results in the literature regarding which β-lg variant is more thermally stable. Our results confirm those reported by Imafidon et al. (*2*), McSwiney et al. (*11*) and Huang et al. (*3, 15*) but contradicted those of Gough and Jenness (*16*), Hillier and Lyster (*17*) and Dannenberg and Kessler (*18*). In a recent study, Nielsen et al. (*19*) showed that the irreversible denaturation of β-lg was concentration dependent. To determine if concentration had an effect on thermal stabilities of the proteins as measured by DSC, the peak temperature of denaturation of the two proteins dispersed in H_2O (pH ~ 6) were measured at different protein concentrations. The results (Table II) showed that between 1 and 10% (w/v) the β-lg B variant was always thermally more stable than the A variant by about 2-3°C.

Electron Microscopy. The transmission electron micrographs of gels formed from β-lg A and B at different pH values are shown in Figures 5 and 6. Marked differences were observed in the microstructure of the gels formed at acid and alkaline pH and also between the microstructures of the gels formed from the A and B variants. Gels made at acidic pH consisted of aggregates of compact spherical globules. Thin sections through these globules revealed a densely staining uniform interior with a fairly smooth boundary (Figures 5a-d). Crude measurements of the globule diameters using a pair of calipers and a graduated rule (an average of 15 globules were measured of several micrographs of each individual gel obtained at the same magnification) suggested that the globules produced by variant A at pH 3.0 and 5.0 (Figure 5a,b) were in general smaller than the globules of variant B (Figure 5c,d). Both variants produced somewhat larger globules at pH 5.0 than at pH 3.0. It should be noted that the protein globules were sectioned at random and the cross sections do not reflect their true dimensions. Statistical corrections would be required to convert such cross sections into true diameter values.

The micrographs further suggested that the compact protein globules formed at acidic pH of 3.0 (Figure 5a and 5c) and 5.0 (Figures 5b and 5d) were either loosely aggregated or fused into chains or clusters of several globules. Aggregation of protein into these structures affected the initially uniform distribution of proteins and created large spaces void of proteins in the gel matrix which are normally filled with the liquid phase of the gel.

The microstructural features of the gels produced under neutral (pH 7.0) and alkaline (pH 8.6) conditions (Figures 6) were completely different from the gels made under acidic conditions. Protein aggregates were much smaller and more evenly distributed in these gels than in the gels made at lower pH. Their clusters were connected through narrow bridges. Scanning electron micrographs (not shown) showed the void spaces between the clusters of these aggregates to be considerably smaller than in the acidic gels which may suggest that the neutral and alkaline gels would have better water holding capacity than the acidic gels.

TEM micrographs of the neutral and alkaline gels (Figure 6) further showed the protein aggregates, particularly those of β-lg A (Figures 6a and 6b) to be fluffy with pores much less than 40 nm in diameter. Porosity was decreased in gels made from β-lg B (Figures 6c and 6d), where the protein clusters were more robust. Whilst the diameter of the spherical aggregates (globules) formed at pH 3.0 and 5.0 ranged from 1 to 2.5 μm, the aggregates formed at pH 7.0 and 8.6 were in the nanometer region.

Table I. Effect of pH on DSC Characteristics of β-lactoglobulin A and B

pH	T_S (°C)		T_D (°C)		T_W (°C)		ΔH (J/g)	
	β-lg A	*β-lg B*	*β-lg- A*	*β-lg B*	*β-lg A*	*β-lg B*	*β-lg A*	*β-lg B*
3.0	79.30	82.35	83.46	85.90	4.00	3.63	12.17	14.80
5.0	78.55	80.80	81.65	83.80	3.25	3.38	10.25	12.65
7.0	69.58	71.07	74.33	75.23	5.00	4.75	9.43	11.30
8.6	66.05	66.58	70.30	71.05	5.63	5.25	6.32	8.80

Table II. Effect of Protein Concentration on DSC Characteristics of β-lactoglobulin A and B

Protein Conc.	T_S (°C)		T_D (°C)	
(w/v)	*β-lg A*	*β-lg B*	*β-lg A*	*β-lg B*
10%	71.34	74.43	77.11	79.45
5%	70.86	73.61	77.28	78.95
2%	69.58	73.54	76.73	79.20
1%	70.05	73.88	77.23	80.04

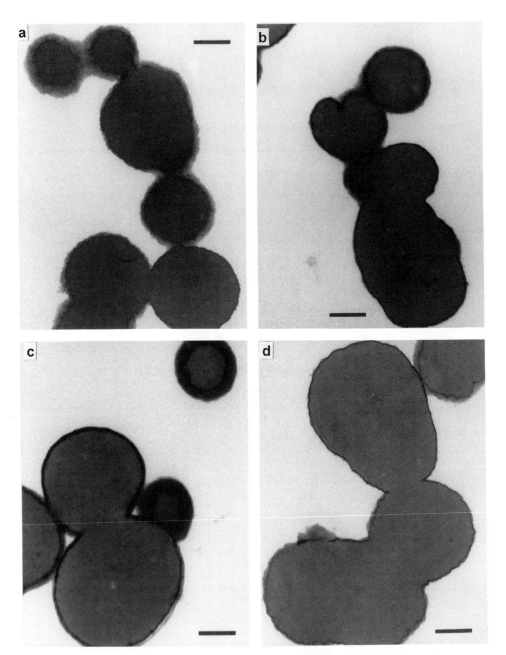

Figure 5. TEM micrographs of β-lactoglobulin A and B gels. [(a) β-lg A, pH 3.0; (b) β-lg A, pH 5.0; © β-lg B, pH 3.0; (d) β-lg B, pH 5.0; Bar represents 0.5μm]. (Adapted from ref. 20).

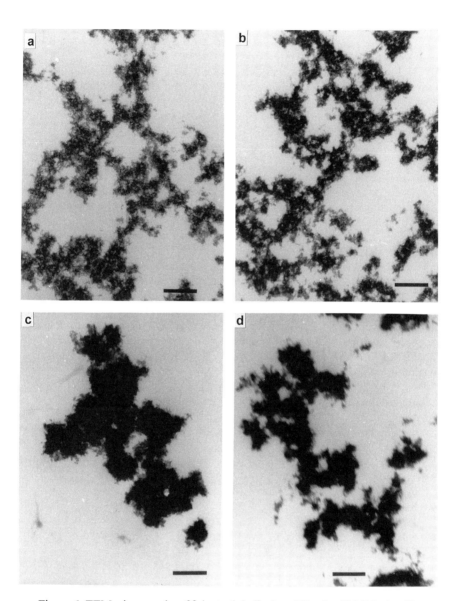

Figure 6. TEM micrographs of β-lactoglobulin A and B gels. [(a) β-lg A, pH 7.0; (b) β-lg A, pH 8.6; © β-lg B, pH 7.0; (d) β-lg B, pH 8.6; Bar represents 0.5μm]. (Adapted from ref. 20).

Furthermore, the sizes of the clusters formed by both variants at pH 7.0 (Figures 6a and c) were different from those formed at pH 8.6 (Figures 6b and d); in addition clusters in the β-lg A gels (at both pH 7.0 and 8.6) were much smaller than in the β-lg B gels resulting in finer strands within the gel network for the A variant when compared with the B variant.

β-lg is an acid stable molecule that unfolds at alkaline pH values (above pH 7.5); decreased charge repulsion at pH 3.0 and 5.0, may have allowed the proteins to form larger more compact globules at acid pH than at alkaline pH. The protein has also been shown to form octamers between pH 3.5 and 5.5 which may further explain the larger aggregates formed at these pH values.

Conclusion

Our results have demonstrated that the A and B genetic variants of β-lg B have different secondary structures at acid pH and relatively similar secondary structure at neutral and alkaline pH. Microstructures of gels formed from the two variants have also been shown to vary. Results from both DSC and FTIR confirm that the two proteins have different thermal stabilities even at protein concentrations as low as (1% w/v). The literature is still conflicting about which β-lg variant has greater thermal stability. Our results indicate that the B variant is more thermally stable. Several studies have indicated that factors such as protein concentration, pH and protein isolation methods can influence denaturation and aggregation. The preparation and storage history of the β-lg samples used in this study from Sigma are not known. Changes may have occurred in their secondary structure during isolation and purification; the results reported in this study should, therefore, be compared with that of β-lg obtained by other separation processes to confirm the findings reported.

Acknowledgments

We would like to thank Drs V. R. Harwalkar and M. Kalab for their valuable contribution. We also thank Miss G. Larocque, Mrs. D. Raymond and Mr. D. Belanger for skilful assistance with electron microscopy and DSC.

References

1. Ferry, J. D. *Adv. Protein Chem.* 1948, *4*, 1.
2. Imafidon, G. I.; Ng-Kwai-Hang, K.F.; Harwalkar, V. R.; Ma, C. -Y. *J. Dairy Sci.* **1991**, *74*, 1791.
3. Huang, X. L.; Catignani, G. L.; Swaisgood, H. E. *J. Agric. Food Chem.* **1994**, *42*, 1276.
4. Kauppinen, J. K.; Moffatt, D. J.; Mantsch, H. H.; Cameron, D. G. *Anal. Chem.* **1981**, *53*, 1454.
5. Allan-Wojtas, P.; Kalab, M. *Milchwissenschaft.* **1984**, *39*, 323.
6. Chirgadze, Y. N.; Fedorov, O. V.; Trushina, N. P. *Biopolymers* **1975**, *14*, 679.
7. Krimm, S.; Bandekar, J. *Adv. Protein Chem.* **1986**, *38*, 181.
8. Susi, H.; Byler, D. M. *Methods for Protein Analysis*; Am. Oil Chem. Soc.: Champaign, IL., 1988, Chapter 14, pp 235.
9. Boye, J. I.; Ismail, A.; Alli, I. *J. Dairy Res.* **1996**, *63*, 97.

10. Clark, A. H.; Saunderson, D. H. P.; Suggett, A. *Int. J. Pept. Protein Res.* **1981**, *17*, 353.
11. McSwiney, M.; Singh, H.; Campanella, O.; Creamer, L.K. *J. Dairy Res.* **1994**, *61*, 221.
12. Kella, N. K. D.; Kinsella. J. E. *Biochem. J.* **1988**, *255*, 113.
13. Privalov, P. L.; Khechinashvili, N. N.; Atanaasov, B.P. *Biopolymers* **1971**, *10*, 1865.
14. Hayakawa, S.; Nakai, S. *J. Food Sci.* **1985**, *50*, 486.
15. Huang, X. L.; Catignani, G. L.; Foegeding, E. A.; Swaisgood, H. E. *J. Agric. Food Chem.* **1994**, *42*, 1064.
16. Gough, P.; Jenness, R. *J. Dairy Sci.* **1962**, *45*, 1033.
17. Hillier, R. M.; Lyster, R. L. J. *J. Dairy Res.* **1979**, *46*, 95.
18. Dannenberg, F.; Kessler, H-G. *J. Food Sci.* **1988**, *53*, 258.
19. Nielsen, B. T.; Singh, H.; Latham, J. M. *Int. Dairy J.* **1996**, *6*, 519.
20. Boye, J. I.; Ma, C. -Y.; Ismail, A.; Harwalkar, V. R.; Kalab, M. *J. Agric. Food Chem.* **1997**, *45*, 1608.

Chapter 12

Limited Proteolysis of α-Lactalbumin and Whey Protein Isolate: Effect on Their Functional Properties

Fakhrieh Vojdani[1] and John R. Whitaker[2]

[1]Department of Agronomy and Range Sciences and [2]Department of Food Science and Technology, University of California, Davis, CA 95616

Millions of tons of whey proteins are produced each year from cows' milk during manufacture of soft and hard cheeses. Some of the whey containing these proteins (β-lactoglobulin, α-lactalbumin, serum albumin, lactoferrin, etc.) is fed to animals, some is discarded and more and more is being used as protein concentrates and isolates in human food. These proteins are relatively soluble except near the isoelectric point (pI) of the proteins, namely pH 4-6. A number of foods, such as fruit juices and vegetable products, have pHs in this same pH region, thereby limiting use of the whey proteins. Perhaps limited proteolysis (1-5%) and/or chemical modifications will increase their solubility, especially near the pI, and may increase their foaming capacity and stability and/or emulsifying activity index and stability. We have shown previously that limited proteolysis (1) and phosphorylation (2) improve the functional properties of β-lactoglobulin.

We report here the effect of limited proteolysis on the functional properties of purified α-lactalbumin and of whey protein isolate from pH 2 to 10 by endoproteinases Glu-C, Lys-C, Arg-C and trypsin.

Materials

α-Lactalbumin. α-Lactalbumin (genetic variant B) was purified from fresh milk of a single cow of the University of California, Davis dairy herd. The milk was cooled immediately and the casein was precipitated by adjusting the pH to 4.6 (1 N HCl added dropwise) within 1 h of milking and the treated sample held at 4°C for 30-60 min (to precipitate all the casein). The supernatant liquid (whey) was separated from the casein by centrifugation (12,000 g for 25 min at 4°C); the fat was removed from the whey solution (top layer). The whey solution was filtered (Whatman #1 paper) dialyzed against distilled water (4°C) frozen, lyophilized and stored at -20°C (whey protein isolate; WPI). All subsequent solutions used for purifying α-lactalbumin used WPI and contained 0.02% sodium azide. α-Lactalbumin was separated initially from the other whey proteins on a Sephadex G-100 column (0.075 M Tris·HCl buffer, pH 7.5) and further purified by DEAE-cellulose anion column chromatography in 0.075 M Tris·HCl buffer, pH 7.5, with a linear gradient of 0.0 to 0.26 M NaCl in the same buffer, followed by CM-cellulose cation column chromatography using 0.01 M sodium acetate buffer, pH 5.0, with a linear gradient from 0 to 0.5 N NaCl in the same buffer. The purified α-lactalbumin gave single bands by polyacrylamide gel electrophoresis (3; data not shown) and by chromatography on a Superose 12 column run on an FPLC system (Figure 1).

Enzymes. The endoproteinases were from Sigma Chemical Co. Endoproteinase Glu-C (*Staphylococcus aureus* V8) has specificity for peptide bonds in which glutamic and aspartic acid residues provide the carbonyl group of the peptide bond. In 50 mM ammonium carbonate and ammonium acetate buffers, hydrolysis of only glutamic acid-containing peptide bonds occurs, while in 50 mM phosphate buffer both glutamic acid and aspartic acid-containing peptide bonds are hydrolyzed. Endoprotease Lys-C (*Lactobacter enzymogens)*) hydrolyzes peptide bonds only where lysine residues provide the carbonyl group. Endoproteinase Arg-C (mouse submaxillary gland) hydrolyzes peptide bonds only where arginine residues provide the carbonyl group. Trypsin (bovine pancreatic) hydrolyzes peptide bonds where arginine and lysine residues provide the carbonyl group. Figure 2 shows the primary sequence of α-lactalbumin variant B and the location of the peptide bonds hydrolyzed by the four endoproteinases. There are three disulfide bonds in α-lactalbumin, which may be broken by reduction before or after proteolysis. In the experiments reported here the disulfide bonds were intact.

Methods

Rates of Hydrolysis. Rates of hydrolysis of α-lactalbumin by endoproteinase Lys-C and by endoproteinase Glu-C are shown in Figure 3. α-Lactalbumin was denatured by heating at 90°C for 30 min at pH 7.0 in 5 mM phosphate prior to enzymatic hydrolysis. The reaction conditions for endoproteinase Glu-C with α-lactalbumin were: 2.0% α-lactalbumin, S/E ratio of 50/1 (enzyme activity of 131.2 units/ml of reaction mixture), pH 7.8 in 0.025 M phosphate buffer containing 0.02% sodium azide, and at 37°C. The conditions were the same for endoproteinase Lys-C, except 469 enzyme units/ml of reaction mixture at pH 8.0 were used. The conditions for endoproteinase Arg-C were as for endoproteinase Glu-C except 38.4 enzyme units/ml of reaction mixture was used, at pH 8.3. Aliquots were removed at various times, the reaction stopped by heating at 98°C and the number of peptide bonds hydrolyzed determined by the 2,4,6-trinitrobenzene sulfonate method (4). Conditions for whey protein isolate hydrolysis with trypsin were the same as for endo-proteinase Lys-C;

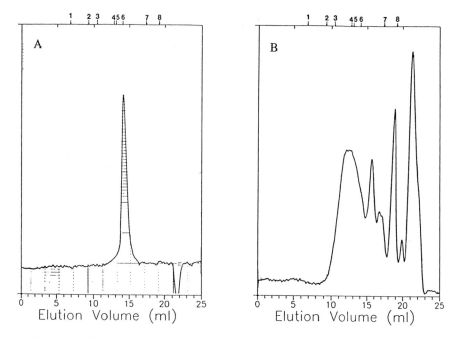

Figure 1. Permeation chromatography of proteins on a Superose 12 column using a Pharmacia FPLC System. A, Purified bovine α-lactalbumin; B, after hydrolysis (DH 4.9) of α-lactalbumin with endoproteinase Glu-C. The numbers at top show elution peaks for: 1, Blue Dextran, two thousand KD; 2, human transferrin (80 KD); 3, bovine serum albumin (66 KD); 4, ovalbumin (43 KD); 5, β-lactoglobulin (18.3 KD); 6, α-lactalbumin (12.4 KD); 7, bradykinin (1.06 KD); 8, prolylglycine (0.172 KD).

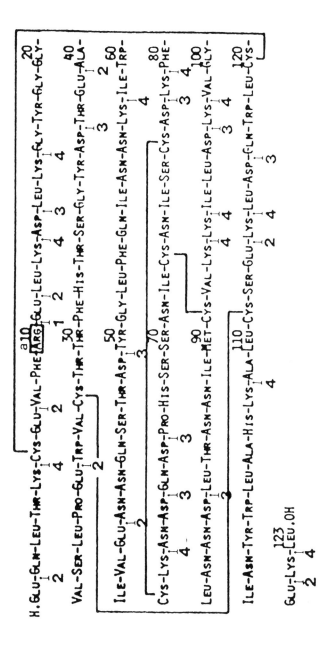

Figure 2. Primary (amino acid) sequence of bovine α-lactalbumin, variant B (11). The numbers indicate peptide bonds susceptible to hydrolysis by: 1, endoproteinase Arg-C; 2 and 3, endoproteinase Glu-C; 4, endoproteinase Lys-C; and 1, 4, trypsin. The long lines indicate disulfide bonds. a, Glycine in variant A.

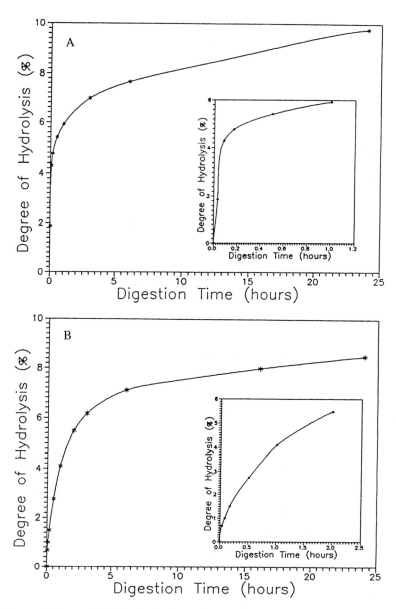

Figure. 3. Rates of hydrolysis of α-lactalbumin by endoproteinase Glu-C (A) and endoproteinase Lys-C (B). The insert magnifies results near beginning. See text for experimental conditions.

2340 enzyme units/ml of reaction mixture was used (ratio of substrate to trypsin was 100/1 (0.01 M phosphate, pH 7.6)). For endoproteinase-C, the conditions were: 2% WPI, 0.1 M phosphate buffer, pH 8.1, 1.0 enzyme units/ml of reaction mixture. The hydrolysis was stopped by addition of TLCK to final concentration of 1.5 mM. The control sample was prepared in the same way as for the proteinase reactions, but no enzyme was added. The control was heated also as for the the enzyme reactions.

Solubilities. The native and modified proteins were weighed and dispersed in distilled water. The pHs of different solutions were adjusted with HCl or NaOH to pH 2 to 10, with constant stirring. The volume was adjusted to give 0.1% protein (w/v). After 1 h equilibration at 23°C, part of the solution was centrifuged at 18,500 x g for 15 min in a microfuge centrifuge. The supernatant was filtered using Microfilterfuge tubes containing 0.2 um Nylon-66 membrane filters. The protein content of the supernatant was determined by the Lowry method (5) using bovine serum albumin as standard.

Emulsifying Activities. The emulsifying activities were determined by the spectroturbidity method of Pearce and Kinsella (6). The Pearce and Kinsella equation was corrected according to Cameron *et al.* (7).

Emulsion Stabilities. Aliquots of the emulsions prepared for emulsifying activities were incubated at 23°C for 24 h in test tubes. After 24 h the tubes were rotated five times to give a homogeneous suspension, an aliquot was removed and diluted 1000 x with 0.1% SDS in 0.1 M NaCl, pH 7.0, to determine turbidity (6) at 500 nm. The remainder of the 24 h old emulsions were heated at 80°C for 30 min, cooled and mixed as above. An aliquot was diluted 1000 x and the turbidity determined. The emulsion stability was calculated as described by Pearce and Kinsella (6).

Foaming Capacity and Stability. The apparatus used for measuring foaming capacity and stability, designed in our laboratory, was made of a calibrated Pyrex glass column (1.0 cm i.d., 41 cm in length) surrounded by a water jacket connected to a pump and water bath at 25.0±0.5°C. The bottom of the column contained a fine fritted glass disc (1.0 cm diameter, 10 C Corning, Ace Glass Co. No. 31001). N_2 gas was bubbled from the bottom of the column at a fixed rate of 30 ml/min to give a uniform, fine N_2 bubble distribution. Increase in foam volume (capacity) was recorded after 30, 40 and 50 sec. Time (sec) required to drain 85 to 92% of the original volume was recorded (stability). Foam capacity was defined as

$$\% \text{ Overrun} = 100(V_t - V_o)/V_o$$

where V_t is volume of foam at any time and V_o is the original liquid volume (2.0 ml).

Results and Discussion

Purified α-Lactalbumin and Hydrolysis Products. The effects of limited, specific hydrolysis of purified α-lactalbumin on solubility, emulsion activity and stability, and foam capacity and stability are demonstrated for endoproteinase Glu-C, endoproteinase Arg-C and endoproteinase Lys-C.

Endoproteinase Glu-C. There are 16 glutamic acid- and aspartic acid-containing peptide bonds in α-lactalbumin (Figure 1). Half of these bonds (7 of 16) are near the N- and C-terminal ends of the molecule, permitting formation of some larger fragments from the middle regions. The disulfide bonds remained intact so that a very large fragment could be produced. The degree of hydrolysis (DH) was 4.9. The molecular weight distribution of fragments that were produced are shown in Figure 1B (compare with Figure 1A for the unhydrolyzed α-lactalbumin).

The functional properties of α-LA-Glu-C (LA-V8; DH 4.9) are shown in Figure 4. The solubility of α-LA-Glu-C was higher than that for the control near the pI region (Figure 4A). This resulted in higher E.A.I. for α-LA-Glu-C at all pHs except at 4 (Figure 4B). However, the emulsion stability was not improved over that of the control (Figure 4C). The foam capacity of α-LA-Glu-C was as good as that of bovine serum albumin at pH 3.5 to 7 and pH 10 (Figure 4D) and the foam stability of α-LA-Glu-C was better than for the control at pHs 4 to 7, with stability at pH 6 being very good (Figure 4E). The solubility and functional properties of native α-lactalbumin (LA) are also shown in Figure 4; however, α-lactalbumin could not be used as the control for α-LA-Glu-C because there was no proteolysis of native α-lactalbumin until it was heat denatured (the control).

Endoproteinase Arg-C. As shown in Figure 1, α-lactalbumin has only one arginine residue. The experimentally determined degree of hydrolysis (DH) was 0.80. The disulfide bonds remained intact so the two peptides formed (R_1-R_{10} and R_{11-123}) remained together. Figure 5A shows that the solubility of the α-LA-Arg-C was less than that of the control at all pHs except for a small improvement at pH 4.6. The emulsifying activity index (E.A.I.) values were all lower for α-LA-Arg-C than for the control (Figure 5B), as were the emulsion stability values (Figure 5C). However, the foam capacity of α-LA-Arg-C (Figure 5D) was better than that for the control at all pHs and even better than that for bovine serum albumin (BSA), a protein that has a high foam capacity (see Figure 4D). The foam stability for α-LA-Arg-C (Figure 5E) was better at all pHs than that for the control. Near the pI of 5, the foam stability was very much better than for the control. Therefore, hydrolysis of one peptide bond (of 122 present) had a marked effect on foam capacity and stability.

Endoproteinase Lys-C. There are 12 lysine residues in α-lactalbumin which endoproteinase Lys-C might hydrolyze (Figure 2). Since the disulfide bonds are intact in α-LA-Lys-C, if all the susceptible peptide bonds were hydrolyzed only 20 of the 123 amino acids would be released as small peptides. Thus, the final product hydrolyzed would have a large fragment with 103 amino acids (11,700 unit weight). The hydrolysate used for functionality had a DH of 4.5.

The functional properties of α-LA-Lys-C (LA-Lys) hydrolysate are shown in Figure 6. The solubility of α-LA-Lys-C hydrolysate was much better at pH 4.5 and

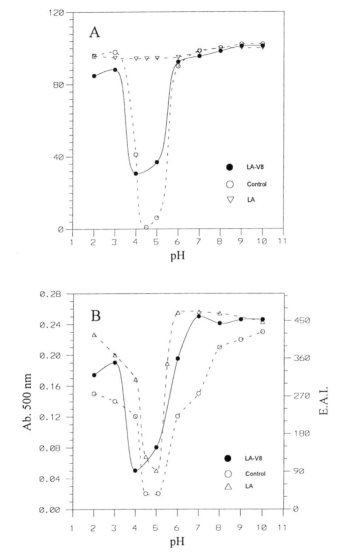

Figure 4. Functional properties of α-lactalbumin before (LA) and after hydrolysis (LA-V8; DH 4.9) by endoproteinase Glu-C. Control was heat treated α-lactalbumin). A, solubility as function of pH. B, Emulsifying activity index (E.A.I.) as function of pH. C. Emulsion stability as function of pH. D. Foam capacity as function of pH. BSA, bovine serum albumin. E. Foam stability as function of pH.

Continued on next page.

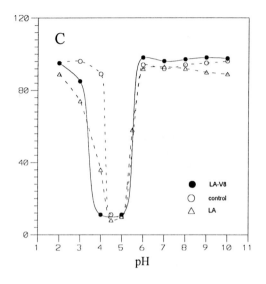

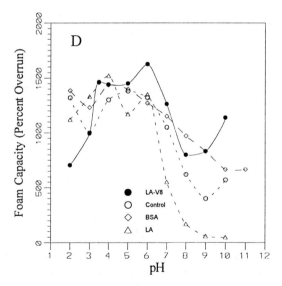

Figure 4. *Continued.*

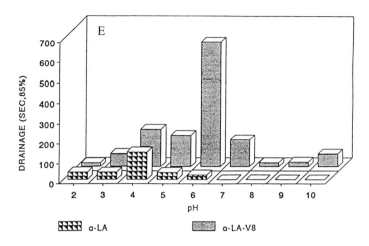

Figure 4. *Continued.*

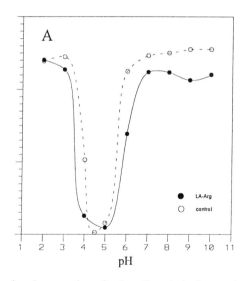

Figure 5. Functional properties of α-lactalbumin before and after hydrolysis (LA-Arg; DH 0.8) by endoproteinase Arg-C. Control was heat treated α-lactalbumin. A, Solubility as function of pH. B, Emulsifying activity index (E.A.I.) as function of pH. C. Emulsion stability as function of pH. D. Foam capacity as function of pH. E. Foam stability as function of pH.

Continued on next page.

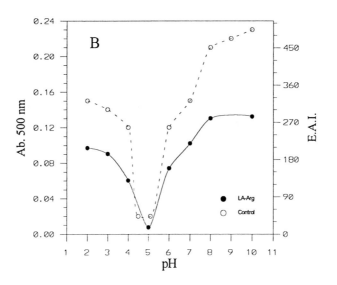

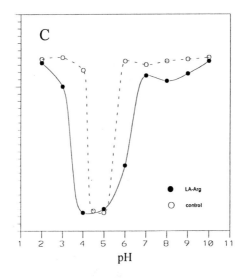

Figure 5. *Continued.*

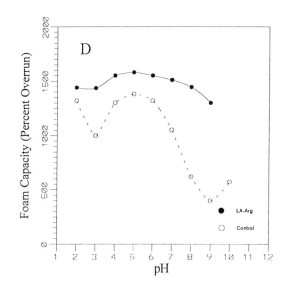

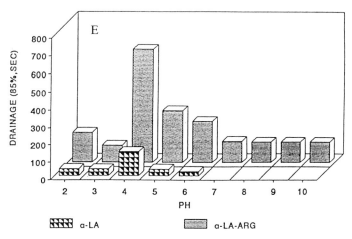

Figure 5. *Continued.*

5 than that of the control (Figure 6A). The E.A.I. of α-LA-Lys-C hydrolysate was better than for the control at pH 5 to 9 (Figure 6B) and the emulsion stability was better at pH 3.5 to 5 and pH 7 to 9 than for the control (Figure 6C). The foam capacity for α-LA-Lys-C hydrolysate was much better than for the control at all pHs and at least as good as that for bovine serum albumin (see Figure 4D). There was marked improvement of the foam stability of α-LA-Lys-C hydrolysate at pH 3 to 6 over that of the control (Figure 6E).

Hydrophobicity of α-Lactalbumin, Its Hydrolysates and Other Proteins. Hydrophobicity was measured fluorometrically, using *cis*-parinaric acid (8) to probe alkyl sites and anilinonaphthalene sulfonate (9) to probe aromatic sites. Table 1 shows the surface and exposed hydrophobicity for α-lactalbumin and its hydrolysates in relation to some other proteins. There are many more alkyl sites (with *cis*-PnA) than aromatic sites (ANS) before (surface) and after (exposed) heat treatment. Before heat treatment, the results show that, with the exception of α-LA-Lys-C (DH 4.5) the hydrolyzed products of α-lactalbumin had more surface hydrophobicity than native α-lactalbumin. This probably is a result of a loss of hydrophobic amino acids, of a total of 40, on hydrolysis and/or a more unfolded nature. α-Lactalbumin has much less surface hydrophobicity than bovine serum albumin and β-lactoglobulin, two highly functional proteins. But α-lactalbumin has more surface hydrophobicity than ovalbumin and trypsin, which have poor functionality.

The calculated hydrophobic sites on the surface of α-lactalbumin and its hydrolysates, using *cis*-parinaric acid, are given in Table II. There were about twice the number of hydrophobic sites for the modified α-lactalbumin derivatives. Bovine serum albumin had 5.00 measured hydrophobic sites, in agreement with literature values (10).

Whey Protein Isolate. The functional properties of whey protein isolate (WPI), trypsin treated WPI (WPI-Try; DH 3.8) and endoproteinase Lys-C treated WPI (WPI-Lys-C; DH 6.5) are shown in Figure 7. The solubilities of WPI and WPI-Try were about the same from pH 2 to 10. WPI-Lys-C had lower solubility at pHs 4 and 5 than WPI; otherwise the solubilities were about the same (Figure 7A) The emulsion activity of WPI-Try immediately after preparation was higher than WPI at all pHs (Figure 7B), and was substantially higher at pHs 2 and 6 to 9. The emulsion activity for WPI-Lys-C immediately after preparation was higher at pH 2 and 6 to 10 than for WPI (Figure 7B). At pH 3-5 it was lower than for WPI. Emulsion activity after 24 h of preparation plus heat for WPI-Try and WPI-Lys-C was similar to emulsion activity at zero time (Figure 7C). The emulsion stabilities were about the same for all three samples (WPI, WPI-Try and WPI-Lys-C) at all pHs except pH 2 and 4 (Figure 7D). At pH 2 WPI-Try and WPI-Lys-C had the same emulsion stability; it was higher than for WPI. At pH 4, WPI-Lys-C had lower emulsion stability than WPI and WPI-Try, which had about the same emulsion stabilities.

The foam capacities for WPI and WPI-Lys-C were essentially the same and independent of pH. WPI-Try had lower foam capacity than WPI at pHs 2 and 3, and somewhat higher values at pHs 5 to 10 (Figure 7E). WPI-Lys-C had very much

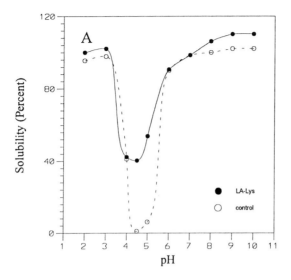

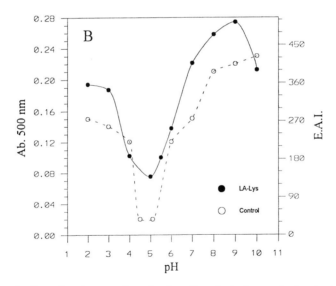

Figure 6. Functional properties of α-lactalbumin before and after hydrolysis (LA-Lys; DH 4.5) by endoproteinase Lys-C. Control was heat treated α-lactalbumin. A. Solubility as function of pH.   B. Emulsifying activity index (E.A.I.) as function of pH.   C. Emulsion stability as function of pH.   D. Foaming capacity as function of pH.   E. Foam stability as function of pH.

Continued on next page.

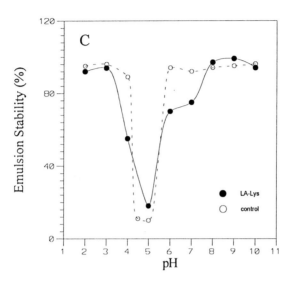

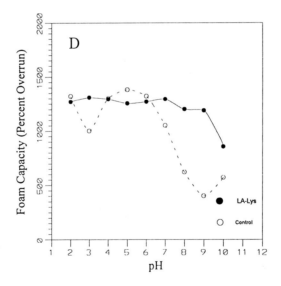

Figure 6. *Continued.*

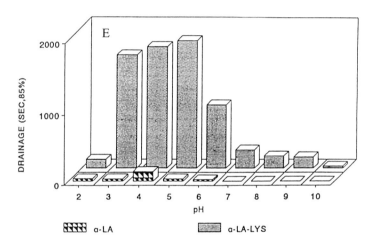

Figure 6. *Continued.*

Table I. Surface and Exposed Hydrophobicity of Proteins by Fluorometric Method Using *cis*-Parinaric Acid and Anilino-naphthalene Sulfonate

Protein	*cis*-PnA		ANS	
	Surface[a]	Exposed[b]	Surface[a]	Exposed[b]
α-La (native)	113	431	8	105
α-La-denatured	162	531	83	214
α-La-Arg-C (DH 0.8)	170	617	109	88
α-La-Glu-C (DH 4.8)	124	314	89	47
α-La-Lys-C (DH 4.5)	27	127	37	37
BSA	1180	-[c]	219	-
Ovalbumin	37	-	-	-
Trypsin	53	-	-	-
β-Lactoglobulin type A	1627	848	54	37
β-Lactoglobulin type AB	1560	1200	58	98

[a]Not heat treated. [b]Heat treated to denature proteins. [c]Not available.

Table II. Hydrophobic Sites on the Surface of Protein Molecules

Protein	Number of sites/molecule
α-Lactalbumin	0.49
Heat denatured a-LA	0.96
α-La-Arg-C (DH 0.8)	1.00
α-La-Lys-C (DH 4.5)	0.95
α-La-Glu-C (DH 4.8)	1.20
Bovine serum albumin	5.00

better foam stability than WPI at all pHs (Figure 7F). WPI-Try had very much better foam stability than WPI at pHs 4 to 10 and about the same at pHs 2 and 3. Only at pH 10 was the foam stability of WPI-Try higher than that for WPI-Lys-C.

Conclusions

Hydrolysis of isolated α-lactalbumin and of whey protein isolate with highly

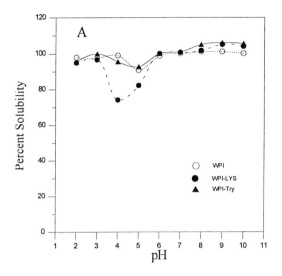

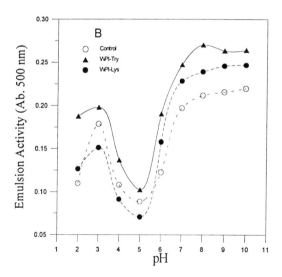

Figure 7. Functional properties of whey protein isolate before (WPI) and after hydrolysis with trypsin (WPI-Try; DH 3.9) and endoproteinase Lys-C (WPI-Lys; DH 6.5). A. Solubility as function of pH. B. Emulsion activity as function of pH immediately after preparation. C. Emulsion activity after 24 h of preparation plus heat as function of pH. D. Emulsion stability as function of pH. E. Foam capacity as function of pH. F. Foam stability as function of pH.

Continued on next page.

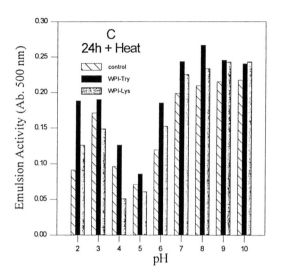

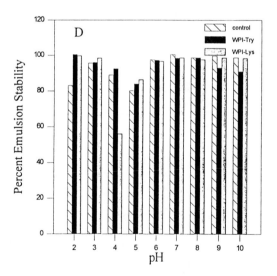

Figure 7. *Continued.*

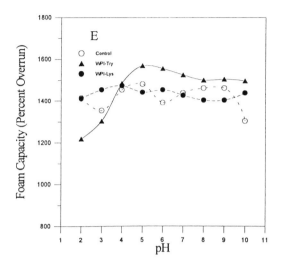

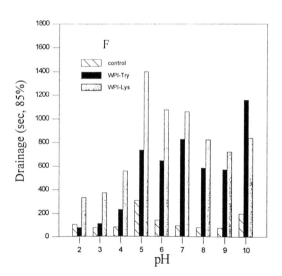

Figure 7. *Continued.*

specific endoproteinases Arg-C, Lys-C, Glu-C and trypsin can be used to improve emulsion activity and stability and/or foaming capacity and stability. Both emulsion and foaming qualities are not improved to the same extent in a single preparation since emulsions and foams depend on different factors for their capacities and stabilities. Foams, air dispersed in liquid, are more dependent on improvement of solubility, especially near the isoelectric point, which is the best pH range for foaming. Emulsions, lipid dispersed in liquid, are dependent more critically on the hydrophobic nature of the system, and a pH that is away from the isoelectric point.

Literature Cited

1. Vojdani, F.; Whitaker, J. R. In *Protein Functionality in Food Systems*; Hettiarachchy, N. S. and Ziegler, G. R., Eds.; Marcel Dekker, Inc.: New York, N. Y. **1994**; pp. 261-309.

2. Vojdani, F.; Whitaker, J. R. In *Macromolecular Interactions in Food Technology*; Parris, N., Kato, A., Creamer, L. K., Pearce, J., Eds. ACS Symposium Series 650; American Chemical Society: Washington, D. C. **1995**; pp. 210-229.

3. Davis, B. J. *Ann. N. Y. Acad. Sci.* **1964**, *121*,404-427.

4. Fields, R. *Methods Enzymol.* **1972**, *25B*,464-468.

5. Lowry, O. H.; Rosebrough, N. J.; Farr, A. L.; Randall, R. J. *J. Biol. Chem.* **1951**, *193*, 265-275.

6. Pearce, K. N; Kinsella, J. E. *J. Agric. Food Chem.* **1978**, *26*, 716-723.

7. Cameron, D. R.; Weber, M. E.; Idziak, E. S.; Neufeld, R. J.; Cooper, D. G. *J. Agric. Food Chem.* **1991**, *39*, 655-659.

8. Kato, A.; Nakai, S. *Biochim. Biophys. Acta* **1980**, *624*, 13-20.

9. Hayakawa, S.; Nakai, S. *J. Food Sci.* **1985**, *50*, 486-491.

10. Sklar, L. A.; Hudson, B. S.; Simoni, R. D. *Biochemistry*, **1977**, *16*, 5100-5108.

11. Whitney, R. M. In *Fundamentals of Dairy Chemistry*; Wong, N. P., Jenness, R., Keeney, M. and Marth, E. H., Eds.; Van Nostrand Reinhold Co.: New York, N. Y., **1988**; pp. 81-169.

Chapter 13

Emulsifying Properties of Cholesterol-Reduced Egg Yolk Low-Density Lipoprotein

Yoshinori Mine and Marie Bergougnoux

Department of Food Science, University of Guelph, Guelph, Ontario N1G 2W1, Canada

The emulsifying properties of cholesterol-reduced egg yolk low-density lipoprotein (CR-LDL) were studied. The CR-LDL was prepared by absorbing cholesterol to β-cyclodextrin (CD). The CR-LDL formed larger emulsion particles at low protein concentrations. Concentration of protein at the interface was greater for emulsions made by CR-LDL when compared to the control LDL at pH 7.0 and 3.5, a result attributed to formation of lipoprotein aggregates by removing the cholesterol in LDL. The emulsion stability of CR-LDL at low protein concentration was lower at pH 7.0 than at pH 3.5, whereas the LDL emulsions showed a considerable stability during aging for one month. Electrophoretic analysis of the adsorbed polypeptides revealed the preferential adsorption of LDL polypeptides at the interface. Increase in protein concentration resulted in a higher phosphatidylethanolamine (PE) and lower phosphatidylcholine (PC) levels at the interface, whereas the opposite trend was observed at pH 3.5. Removing the cholesterol from egg yolk LDL caused changes in phospholipids-protein interactions at the interface which could be explained the instability of CR-LDL emulsion.

Hen's egg yolk provides excellent functional properties to a variety of food products such as mayonnaise, ice cream, bakery items and salad dressings. Egg yolk contains various emulsifying agents such as hydrophobic and hydrophilic proteins, phospholipids and cholesterol. Egg yolk consists of soluble plasma (-78% of the total liquid yolk) which is composed of livetins and low-density lipoprotein (LDL) *(1)*. LDL contains -12.5% protein and -80% lipids. The lipid in LDL consists of 70% neutral lipid, 26% phospholipids (71-76%, phosphatidylcholine (PC), 16-20%

phosphatidylethanolamine (PE), and 8-9% sphingomyelin and lysophospholipids), and 4% free cholesterol *(2)*. LDL has been considered the major factor governing the emulsifying properties of egg yolk. Protein-phospholipids complexes (lipoproteins) are the components of egg yolk responsible for stabilizing an emulsion *(3,4)*. In an emulsion prepared with egg yolk, the contribution of proteins to emulsifying activity is higher than that of phospholipids *(5)*. Egg yolk proteins exhibit a higher adsorbing capacity than globular proteins, because they have a more flexible structure and a greater surface hydrophobicity *(6)*. The emulsifying properties and heat stability of protein emulsion were improved substantially through the formation of a complex between lysolecithin and free fatty acids *(7,8)*. Emulsifying capacity and heat stability of egg yolk were also improved by fermentation with pancreatic phospholipase *(9)*. These results indicate that the emulsifying properties of egg yolk lipoproteins might be closely related to the structure of phospholipids-protein complexes and their interactions at an oil-in-water interface. On the other hand, concerns regarding the relationship between cholesterol or oxidized cholesterol products and coronary heart disease attempts have resulted in various technologies being examined to reduce the cholesterol content in egg yolk *(10)*. These include: solvent extraction, supercritical fluid extraction, enzymatic degradation and complexing with β-cyclodextrin (CD)*(11)*. The removal of cholesterol by adsorption to CD is an alternative approach. A few papers have reported on the functional properties of low-cholesterol egg yolk *(5, 12, 13)*. However, no information is available on adsorption behavior and the effect of cholesterol reduction from egg yolk on its phospholipids-apoprotein interactions at an oil-in-water interface. The objective of this chapter is to present emulsifying properties of cholesterol-reduced LDL (CR-LDL) and phospholipids-protein interactions at the interface.

Materials and Methods

Preparation of Cholesterol-Reduced Egg Yolk Low-density Lipoprotein (CR-LDL) LDL was prepared from fresh egg according to the modified method of Raju and Mahadevan *(14)*. LDL concentration was determined from protein concentration using a modified Lowry procedure *(15)*. Extraction of cholesterol from LDL was carried out using CD. The LDL solution (6.0%, w/v) was heated to 50°C in a water bath and CD (Wacker Chemicals, CT) was added at a CD:cholesterol molar ratio of 2 and 4. The sample was mixed for 45 min at 50°C and cooled at 4°C for 1 hrs. The slurry was centrifuged for 30 min at 8000g at 10°C. The supernatant containing the CR-LDL was decanted and used for the preparation of emulsions. The determination of egg yolk composition and the reduction ratio of cholesterol in CR-LDL were measured using flame ionization (TLC-FID) on an Iatroscan (Iatroscan MK-5, Iatron Laboratories, Inc, Tokyo, Japan).

Determination of Emulsifying Properties The LDL and CR-LDL preparations were diluted with various buffers (50mM acetate and imidazole buffers containing 0.1M and 1.5M NaCl, pH 3.5 and 7.0, respectively) to give a final LDL concentration 0.4-4.0 % (w/v) in the aqueous phase. Emulsions were prepared by homogenizing 2.0 mL for each LDL or CR-LDL solution with 0.5 mL of pure triolein (>99%) for 1 min at a speed of 22,000 rpm using a Polytron PT 2000 homogenizer (Kinematica AG, Switzerland). The droplet size distribution of emulsions was determined on a Mastersizer X (Malvern Instruments Ltd., Malvern, UK) with optical parameters defined by the manufacture's presentation code 0303. The emulsions were centrifuged at 20°C and 5000g for 30 min and the cream washed with 5 mL of appropriate buffer with each washing followed by centrifugation. The subnatants were pooled together, filtered through a 0.22µm filter. The protein contents were determined according to the modified Lowry method *(15)*. The surface concentration was estimated as a difference between protein concentration of the subnatant solution and the total protein used to make the emulsion. The washed cream was treated with a solution of 10% (w/v) SDS in 0.1M Tris buffer, pH 8.0, containing 5% (v/v) 2-mercaptoethanol (ME). The protein composition of the supernatant was determined by SDS-polyacrylamide gel electrophoresis (SDS-PAGE) on 4-15% gradient gels using the Bio-Rad Mini Protean II electrophoresis cell at a constant voltage of 20 mA/gel.

Analysis of Phospholipids and Cholesterol Present at the Interface
The creams were extracted with approximately 5 volumes of chloroform:methanol (2:1, v/v) and solvent layer was evaporated using a rotary evaporator, the residue dissolved in 5 ml of 5% (v/v) ethyl acetate in hexane and loaded onto a prepacked Sep-Pack silica cartridge (Waters Co., Milford, MA). The column had been previously dehydrated in succession with 5 ml ethyl acetate, 10 ml 50% (v/v) ethyl acetate in acetone, 5 ml ethyl acetate and 20 ml hexane. The unadsorbed lipid (triolein) were washed off the column with 20 ml of 5% (v/v) ethyl acetate in hexane and a portion of the eluate collected and analyzed as described below for phospholipids and cholesterol. The phospholipids and cholesterol were then eluted from the column with 20 ml of methanol:water (98:2, v/v) and the eluate collected and transferred into evaporation flasks. The contents of the flasks were evaporated to dryness on a rotary evaporator and the residue dissolved in 0.5 mL of chloroform:methanol (2:1, v/v) solvent. The phospholipids contents were then analysed by TLC-FID using the Iatroscan system. The samples were developed for FID as follows: one µL of sample was applied onto the Chromarods (Chromarod-S3), dried and developed first in chloroform-methanol-water (70:30:3, v:v:v) for 10 min. The Chromarods were dried and developed in a second solvent system containing petroleum ether-diethyl ether-acetic acid (80:30:0.2, v:v:v) for 30 min. After drying, the Chromarods were loaded onto the Iatroscan system and the area under each peak determined and used to calculate the concentration of PC, PE and cholesterol. Egg PC (QP Corporation, Tokyo, Japan), cholesterol (Sigma Chemicals) and bovine liver PE (Avanti Polar Lipids Inc., CA) were used as standards.

Results and Discussion

Cholesterol Reduction from Egg Yolk LDL. Table 1 shows the effect of CD: cholesterol molar ratio on the extraction of cholesterol from egg yolk LDL. The reduction of cholesterol in LDL was 48.5 and 92.7 % for molar ratios of 2 and 4 CD: cholesterol. The removal of cholesterol from liquid egg yolk by adsorption to CD has been described by several researchers *(16)*. The most important factors influencing cholesterol reduction were dilution of egg yolk to a defined water : solid ratio (2.9) and CD concentration at a CD: cholesterol molar ratio of 4.0 *(16)*. CD remaining in the sample after centrifugation was determined using a freshly prepared phenolphthalein solution *(16)*. The value for each sample was negligible.

Table 1. Effect of β-cyclodextrin (CD): cholesterol molar ratio on the reduction of cholesterol from egg yolk LDL.

Samples	*Ratio of CD; cholesterol (molar ratio)*	Cholesterol reduction (%)	Residual CD (%)
CR-LDL (1)	2	48.5	0.025
CR-LDL (2)	4	92.7	0.074

Emulsifying Properties of Cholesterol-Reduced LDL. The change in particle size of the emulsions as a function of LDL concentration and pH are shown in Figure 1. The mean particle size of emulsions decreased with increasing LDL concentration used to make the emulsion.   Generally, there was a decrease in particle size with increasing amount of emulsifier.   The latter observation is supported by the fact that an increase in the concentration of surface active agents generally leads to a reduction in interfacial surface tension of the droplets, facilitating their break-up into smaller droplets *(17)*. At low surface concentrations the system produces larger emulsion droplets since there is insufficient surfactant to cover all of the freshly created oil surface so that droplet size increased. At high surfactant concentration, the system produces maximal interfacial area under the proceeding thermodynamic conditions. At concentrations of LDL above 2.4%, the droplet sizes of emulsions plateaued at 0.8-0.9 µm. There was no noticeable change in particle size for pH 7.0 and 3.5. The CR-LDL formed larger emulsion droplets than LDL at 0.4 % at pH 7.0 and 3.5.   The results would indicate that LDL is a better emulsifier at low concentrations than CR-LDL at neutral and acid pH values. Figure 2 shows the amount of protein present on the surface of the emulsion as a function of LDL concentration.   At pH 7.0, the surface protein concentration of LDL emulsion increased from 0.24 to 1.25 mg/m^2 with increasing LDL concentration. The concentration of protein at the interface was greater for emulsions stabilized with CR-LDL at pH 7.0.   However, there was no noticeable difference between 48.5% CR-LDL and 92.7% CR-LDL. On the other hand, the surface protein concentration was markedly different at pH 3.5.   The LDL surface protein ranged from 0.32 to 1.08 mg/m^2.   However, the 48.5% CR-LDL and 92.7% CR-LDL formed much thicker films, ranging from 0.67-1.56 and 0.74-3.01 mg/m^2, respectively.   These data suggest that the adsorption behavior of CR-LDL at an oil-in-water interface is different from LDL at pH 3.5.   The higher surface

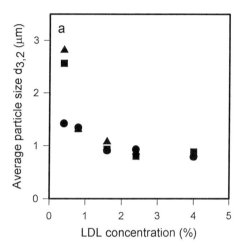

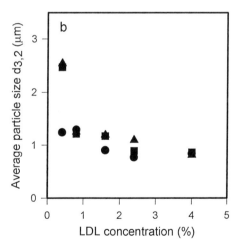

Figure 1   Mean droplet size of emulsions (20% oil, v/v) stabilized by egg yolk LDL
(●), 48% CR-LDL (■) and 93% CR-LDL (▲) as a function of LDL
concentration at pH 7.0 (a) and 3.5 (b).

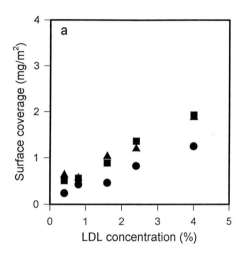

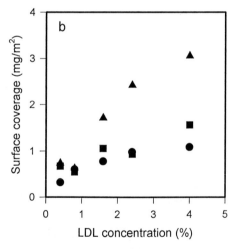

Figure 2   Surface protein coverage of emulsions (20% oil, v/v) stabilized by egg yolk LDL (●), 48% CR-LDL (■) and 93% CR-LDL (▲) as a function of LDL concentration at pH 7.0 (a) and 3.5 (b).

protein concentration of 92.7% CR-LDL at pH 3.5 when compared to the control LDL can be explained on the basis of increased coagulation of lipoproteins at low pH values. More recently, we reported that egg yolk LDL micelles breakdown when the micelles come into contact with the interface and rearrangement of lipoproteins, cholesterol and phospholipids occurs following adsorption at an oil-in-water interface *(18)*. The ^{31}P NMR and enzymatic hydrolysis studies showed that the membrane fluidity of egg yolk LDL is high and the interactions of protein-phospholipids may not be so strong *(19)* as proposed by Burley *(20)*. There is no information regarding the role of cholesterol on structural change of egg yolk LDL. In general, it is believed that cholesterol is an important component in stabilizing biological cell membranes. Removing the cholesterol from egg yolk LDL may cause the structural change of phospholipids-protein interaction in LDL micelles. These changes may be the major factor for forming thicker film when cholesterol comes to the interface.

Protein Composition at the Oil-in-water Interface It has been reported that LDL consists of about six major polypeptides that range in molecular mass from about 10 to 180 kDa and several unidentified minor polypeptides. *(14, 21)*. The SDS-PAGE analysis indicated LDL was composed of 9 major polypeptides ranging from 19-225 kDa, and some minor polypeptides. From the migration patterns of polypeptides in SDS-PAGE gels (Figure. 3), the preferential adsorption was observed among polypeptides in LDL emulsions. Almost all major polypeptides with high molecular mass (> 60 kDa) in LDL components were adsorbed on the oil surface, but four major polypeptides with molecular mass (< 48 kDa) remained in serum. Such results were obtained with CR-LDL emulsions at LDL concentrations ranging from 0.4 to 4.0%. Even at the lower LDL concentration (0.8%), these four polypeptides did not adsorb at an O/W interface. The molecular sizes of these unabsorbed polypeptides were estimated at 48, 43, 40 and 19 kDa, respectively.

Phospholipids and Cholesterol Composition at the Interface The compositions of phospholipids and cholesterol at the oil-in-water interfaces were shown in Figure 4. The emulsifying properties of egg yolk LDL have been attributed in part to the phospholipids-protein complex which can interact with the oil phase through hydrophobic groups and also with the aqueous phase through the charged phospholipid groups. The percentage of phospholipids and cholesterol in the LDL was PC, 66.6%, PE, 19.0% and cholesterol, 16.4%. These results were similar to previous reported values *(2)*. At pH 7.0, PC level at the interface decreased while that of PE increased with increasing LDL concentration, whereas the opposite trend was observed at pH 3.5. These data indicate that PC is preferentially bound to the interface at pH 7.0 and low protein concentrations when compared to pH 3.5. The results can be explained on the basis of differences in affinities of the PC and PE molecules with apoproteins at different pHs. It is known that the quaternary head group of PC is a stronger base than the primary head group of PE. Therefore, at pH 7.0, PC has more

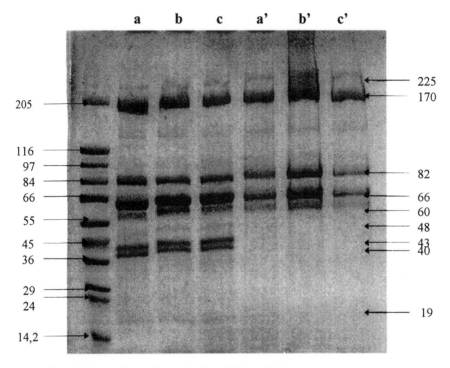

Figure 3   SDS-PAGE profiles of egg yolk lipoproteins.
Lane a a': LDL, Lane b, b': 48% CR-LDL, Lane c, c': 93% CR-LDL
Lane a-c: control, Lane a'-c': emulsions stabilized by 24 mg/mL of LDL or
CR-LDLs.

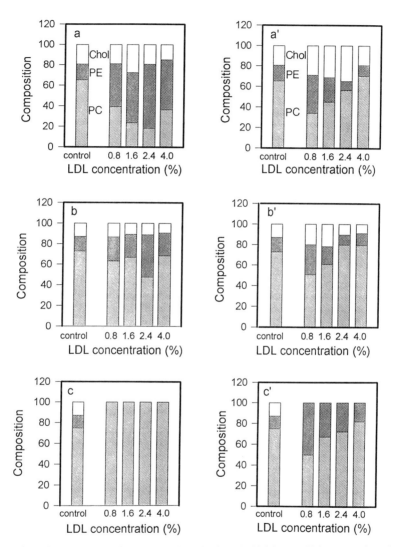

Figure 4   Composition of cholesterol and phospholipids at oil-in-water interfaces stabilized by egg yolk LDL (a, a'), 48% CR-LDL (b, b') and 93% CR-LDL (c, c') as a function of LDL concentration at pH 7.0 (a-c) and 3.5 (a'-c').

▨ : PC, ▨ PE, ▢ cholesterol

charges on the head group, the electrostatic interactions between the PC molecules and apoprotein is weaker than the PE-apoprotein interaction. For the low LDL concentration, the insufficient apoprotein in LDL can not cover all of the freshly created oil surface, so that the PC can adsorb tightly. The proportion of PE increased with increasing protein concentration. However, the negative charges on PC at pH 3.5 were reduced, and the electrostatic interaction between PC molecules and apoproteins become greater as compared to results seen at pH 7.0. Such increased interactions would enable the PC molecule to bind more to the interface with increasing protein concentration at pH 3.5, when compared to the weakly primary base head group of PE *(22)*. Interestingly, the ratio of PC increased and that of PE decreased at 4.0% of LDL at pH 7.0, unlike the trend observed at pH 3.5. As described above, the plateau was reached for particle size at 2.4 % LDL. The decreased cholesterol level at higher protein concentration may be the result of competitive adsorption by the apoprotein and phospholipids. Cholesterol showed less affinity to the interface at pH 7.0 compared to pH 3.5. The cholesterol level at the interface increased with increasing LDL concentration at pH 7.0 and 3.5, while it decreased at 4.0% LDL. Cholesterol is a hydrophobic lipid. At low pH, the interaction between cholesterol and phospholipids or apoprotein could increase. For 92.7% CR-LDL emulsions, very different phospholipids compositions were observed at different pHs. The PE molecule could not be adsorbed to the interface at pH 7.0, while it was adsorbed at the interface at pH 3.5. The level of PE decreased and that of the PC increased with increasing CR-LDL concentration similar to LDL emulsions. The lipid composition from 48.5% CR-LDL emulsion showed intermediate patterns at both pHs for both LDL and 92.7% CR-LDL sample. The interactions of cholesterol with apoprotein or phospholipids are not well understood. The results indicate that cholesterol in LDL would play an important role as a "bridge" to facilitate the interaction of PE at the interface. However, PE can penetrate at the interface at low pHs because of reduced electrostatic interaction with the PC and apoproteins. The differences in lipids composition at the interface can be related to the characterization of LDL and CR-LDL emulsions, such as stability.

Stability of LDL and CR-LDL Emulsions   The stability of the emulsions containing 0.8 and 4.0 % LDL or CR-LDLs at different conditions are shown in Figure 5. Regardless of pH and NaCl concentration, the emulsions stabilized with high LDL concentration were stable up to 1 month. At pH 7.0 and 0.1M NaCl, the CR-LDLs were unstable after 2 weeks and the mean particle size of the emulsion increased with increasing aging time. However, the CR-LDL emulsions were more stable at high NaCl concentration at pH 7.0 for up to 2 weeks. Interestingly, CR-LDL emulsions were stable at pH 3.5 and 0.1M NaCl with low surface concentrations, while coalescence/flocculation of the emulsions increased at 1.5M NaCl, resulting in an increase in emulsion particle size. It has been stated that emulsions stabilized with proteins should be more unstable at low pHs, however, the present results do not support this statement. The lipid

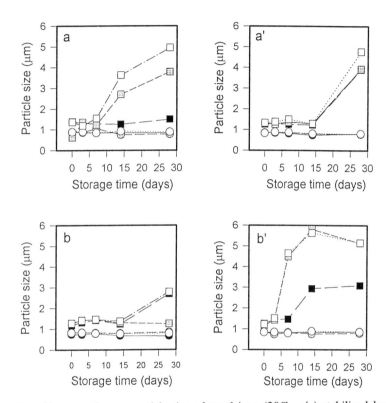

Figure 5   Changes of mean particle size of emulsions (20%, v/v) stabilized by egg yolk LDL (●, ■), 48% CR-LDL (○, ❑) and 93% CR-LDL (○, ❑) as a function of storage time at pH 7.0 (a,a') and 3.5 (b,b') containing 0.1M NaCl (a,b) and 1.5M NaCl (a', b').
(●, ○, ○): 40 mg/mL of LDL or CR-LDL concentration. (■, ❑, ❑): 8 mg/mL of LDL or CR-LDL concentration

composition was also analyzed for emulsions which had undergone 3 weeks of aging. The PC was more dissociated from the interface of CR-LDL emulsions when compared to the control LDL emulsions. The PC molecules were retained on the emulsions at pH 7.0 and 1.5 M NaCl, whereas they were dissociated from the interface at pH 3.5 and 1.5 M NaCl. From the SDS-PAGE results, polymerization of the adsorbed protein was observed during aging (data not shown). In this respect, the emulsion stability of LDL is closely related to phospholipids-apoprotein interaction at the interface. The results indicate that the motional freedom of PC molecules is affected by the interaction of PE, cholesterol and apoprotein. The pHs and salt concentration are also important factors affecting the phospholipids-protein interaction at the interface. At pH 7.0 and low NaCl concentration, the motional freedom of PC molecules of CR-LDL emulsions are relatively high and may be easily dissociated from the interface during aging. On the other hand, the charge of PC is neutralized at high NaCl concentration or low pH values, resulting in a decrease in the dissociation of PC from the interface. The PE molecules at the interface may be an important molecule by acting as a bridge between the PC and oil phase or apoproteins. At pH 3.5 and 0.1M NaCl, the results were similar to those at pH 7.0 and 1.5M NaCl which showed that lowering the pH reduces the charge of PC and apoproteins, whereas the presence of high salts at low pHs may decrease the binding affinity of the phospholipids at the interface, resulting in the breakdown of emulsion droplets.

Conclusion

The effect of cholesterol reduction from egg yolk LDL on its emulsifying properties was investigated. Our results demonstrate that cholesterol is an important component in the stabilization of LDL emulsions. Removing cholesterol from LDL caused the formation of larger particle sizes at the low protein concentration and also formed much thicker films at the interface as a result of apoprotein aggregates. It was also observed that removing cholesterol from LDL changed the phospholipids-apoprotein interactions at the interface and these changes may be responsible for the instability of CR-LDL emulsions. Further studies on phospholipids-protein interactions related to structure-function relationship would be useful for better understanding of egg yolk lipoprotein functionalities.

Acknowledgments

This research was supported by the Ontario Egg Producer's Marketing Board (Ontario, Canada) and the Natural Sciences and Engineering Research Council of Canada (*NSERC*).

Literature Cited
1. McCully, K.A.; Mok, C.C.; Common, R.H. *Can. J. Biochem. Physiol.* **1962**, *40*, 937-952.
2. Martin, W.G.; Tattrie, W.G.; Cook, W.H. *Can. J. Biochem. Physiol.* **1963**, *41*, 657-666.
3. Vincent, R.; Powrie, W.D.; Fennema, O. *J. Food Sci.* **1966**, *31*, 643-647.
4. Mizutani, R.; Nakamura, R. *Lebensm.-Wiss. Technol.* **1985**, *18*, 60-63.
5. Bringe, N.A.; Howard, D.B.; Clark, D.R. *J. Food Sci.* **1996**, *61*, 19-24.
6. Kiosseoglou, V.D.; Sherman, P. *Colloid & Polymer Sci.* **1983**, *26*, 502-507.
7. Mine, Y.; Kobayashi, H.; Chiba, K.; Tada, M. *J. Agric. Food Chem.* **1992**, *40*, 1111-1115.
8. Mine, Y.; Chiba, K.; Tada, M. *J. Agric. Food Chem.* **1993**, *41*, 157-161.
9. Dutilh, C.E.; Groger, W. *J. Sci. Food Agric.* **1981**, *32*, 451-458.
10. Froning, G.W. In *Egg Science and Technology 4th edition*, Stedelman, W.J. Cotterill, O.J. Eds.; Food Products Press; Binghamton, N.Y., **1994**.
11. Merchant, Z.M.; Anilkumar, G.G.; Krishnamurthy, R.G. **1991**, U.S. patent 5037661
12. Awad, A.G.; Bennink, M.R.; Smith, D.M. Poultry Sci. **1997**, *76*, 649-653.
13. Paraskevopoulou, A.; Kisseoglou, V. *Food Hydrocolloids* **1995**, *9*, 205-209.
14. Raju, K.S.; Mahadevan, S. *Anal Biochem.* **1974**, *61*, 538-547.
15. Markwell, M.A.K.; Haas, S.M.; Bieber, L.L.; Tolbert, N.E. *Anal. Biochem.* **1978**, *87*, 206-210.
16. Smith, D.M.; Awad, A.C.; Bennink, M.R.; Gill, J.L. *J. Food Sci.* **1995**, *60*, 691-694, 720.
17. Parker, N.S. *CRC Crit. Rev. Food Sci. Nutr.* **1987**, *25*, 285-315.
18. Mine, Y. *J. Agric. Food Chem.* **1997**, *46*, 36-41.
19. Mine, Y. *J. Agric. Food Chem.* **1998**, *45*, 4564-4570.
20. Burley, R.W. *CSIRO Food Res. Q.* **1975**, *35*, 1-5.
21. Burley, R.W.; Sleigh, R.W. Aust. *J. Biol. Sci.* **1980**, *33*, 255-268.
22. Scholfield, C.R. In *Lecithins-Sources, Manufacture & Uses.* Szuhaj, B.F. Ed., AOCS, Champaign, IL, **1989**.

Chapter 14

Functional Properties of Goat Meat Proteins

A. Totosaus, I. Guerrero, and P. Lara

Departamento de Biotecnología, Universidad Autónoma Metropolitana, Apartado
Postal 55-535, Mexico D. F., C. P. 09340, México

Protein functionality can be classified into three groups: related to
hydration, to protein-protein interactions and to surface properties.
In meat systems, sarcoplasmic proteins show almost no functional
properties whereas the structural arrangement of myofibrillar
proteins are the main cause of their functionality. This is due to the
effect form and molecular flexibility have on hydrodynamic and
surface properties related to hydrophobicity, electrostatic forces and
steric hindrance. To study functionality, the use of a model system
of simple composition and process variables easy to control is more
adequate than the use of real systems. In the search of an alternative
protein source, some functional properties of goat meat extracts
stored at different times and temperatures were studied and
compared to beef extracts. Only the emulsifying activity index had a
highly significant difference with respect to animal species and
storage time. Emulsifying capacity and sulfhydryl group content
were also significantly affected by storage time. Goat protein
extracts showed similar functionality to beef protein extracts.

Functional Properties

Factors influencing protein functional properties are based on physicochemical
characteristics such as amino acids composition, molecular conformation and size,
and chemical bonds and charges involved in their folding. In this way, processes
applied including protein extraction method, temperature, ionic strength and
storage, among others, in addition of physical, chemical or biological modifications
affect protein functionality (1).

Functional properties in a biomolecule are related to non-nutritional properties
than determine the use, either as process aid or as a food ingredient due to their
attributes in which their behavior during processing, storage, preparation and
consumption is based on physicochemical characteristics of the system (2-5).

These properties can be classified into three groups: a) those depending of
protein-water interaction or hydration properties, such as solubility; b) those

depending on protein-protein interactions, such as formation of a network during aggregation; and c) surface properties, depending on surface tension, such as foam and emulsion formation (6). All these three groups of characteristics result from two molecular attributes: hydrodynamics affected by molecule shape and flexibility, and surface-related attributes where hydrophobicity, hydrophility, electrostatic forces and steric hindrance determine the behavior between two phases (7).

Protein functionality is of particular importance to food characteristics such as: 1) sensory, where they are responsible of the color, flavor or sweetening; 2) those maintaining specific characteristics in formulations, such as foaming, emulsifying or water-binding; and, 3) processing, related to flow characteristics such as viscosity or emulsification (8).

Muscle Proteins.   Total protein content in striated muscle varies between 18-20%. It can be divided in three groups: sarcoplasmic proteins, soluble in low ionic strength solutions; myofibrillar or structural proteins, soluble in high ionic strength solutions and connective tissue proteins, insoluble in salt solutions. Myofibrillar proteins are mainly actin, myosin and actomyosin, their isoelectric point is between 4.5 and 5.5, Sarcoplasmic proteins: myoglobin, myogenin, hemoglobin and albumin among others, are mostly related to metabolic activity, with isoelectric point between 6.0-7.0 (9-11).   Proteins of animal origin having the highest functional properties are those included in the myofibrillar system (13). In pre-rigor muscle, myosin is generally considered the mayor contributor to functionality, whereas in post-rigor meat actomyosin fulfills this role (5).

Muscle proteins participate in three kinds of interactions: protein-water, protein-lipid and protein-protein, these basic interactions are characterized by water-binding, fat-binding and gelification properties, respectively (3). In this respect, functional characteristics of meat proteins are related to muscle changes depending on pre-rigor conditions of the animal such as age, sex, nutrition and general handling, as well as slaughtering and storage conditions, until processing or consumption (13).

Meat Protein Solubility.   Solubility can be defined as total percentage of protein remaining in solution under moderated centrifugal forces (14). In this way solubility represents saturation at equilibrium between a solute (protein) and a solvent (water or salt solution). Meat product fabrication is based on protein solubilization in salt solutions (15). Other protein transformations such as gelification and emulsification are affected directly by solubility (2, 9). In meat batters, a cohesive matrix is formed by a protein network. The amount and characteristics of the extracted proteins depend on previous conditions, such as final pH of the muscle and ionic strength of the extracting solution (16). High solubility is desirable due that it allows fast protein dispersion in large quantities, resulting in an efficient diffusion towards water/air or water/oil interfaces. Environmental conditions, as pH, temperature, ionic strength and interactions with other meat constituents, had also great influence on solubility (17).

Meat Emulsions. A meat emulsion is a two phases system, made by a dispersed fatty solid in a continuous phase of water and various soluble components, as salts and proteins, therefore forming a multiphasic system (18). Dispersion is achieved by applying a shear force, however, as the system is thermodinamically unstable, an emulsifying agent is required to stabilize the emulsion (9). These type of emulsions are more stable if the surface tension is low, but as sodium chloride is present and it increases water surface and interfacial tension, it results in a thermodynamically unstable system (18). A meat emulsion can be considered an emulsion like-gel in which fat is dispersed uniformly in a thermally formed continuos protein matrix. This gel-like emulsion differs from an oil/water emulsion in their physicochemical attributes due that the interfacial film, which plays a mayor role in the stabilization of fat globules, always remain is a state of suspension (16, 19, 20).

Comminuted meat systems are not true "meat emulsions", referred in the classic emulsion definition and the term "meat batters" is perhaps more appropriated to describe the multiphase system in which a continuos hydrophilic matrix of salt-water-protein has the role of chemical and mechanical stabilizer in the phase dispersion of raw and cooked products (14). In emulsified meat products meat proteins, specifically those soluble in high ionic strength solutions, are the main emulsifying agents. This proteins, corresponding to the myofibrillar system, dissolve in the aqueous phase and cover the fat forming a protein surface. Myofibrillar proteins had the most efficient functional properties in a meat system and help to stabilize the emulsion in a larger extent than sarcoplasmic proteins (21-23). In addition, as sarcoplasmic proteins are related to several enzymatic systems, their presence could cause destabilization of the system due to hydrolysis (24). The characteristics of meat proteins, expressed in terms of solubility, percentage of hydration and emulsifying capacity depends of factors such as animal species, sex, age, pre and post-mortem conditions, handling, pH and brine concentration in case of fabricated products (25). Salt soluble proteins are liberated by meat mincing and salt addition during the emulsifying process (3) when 84% of the original protein is extracted by the brine (26).

In order to evaluate functional properties of meat proteins model systems are often used. The simple composition, easy handling, less working volumes and process control allow to understand physicochemical and thermodynamic performance of the main meat product ingredients: myofibrillar proteins. The functional parameters more often reported are: emulsifying activity index which describes the total surface area created for a fixed protein and oil concentrations, it is reported in m^2/g or in turbidity units; emulsion stability index describing the time necessary for an emulsion to collapse or aggregate into oil droplets, it is reported in h; and emulsifying capacity, the amount of emulsified oil for a given protein concentration, reported in g oil emulsified before emulsion disruption.

Goats As Meat Producers

The importance of goats as meat-producing animals varies worldwide. As goats are used primarily for textile and milk industries, meat is considered a secondary product (27). However, goats present various advantages as meat producing animals compared to others species: their high reproductive rate, short interval between pregnancies, resistance to harsh climates and small carcasses, easy to preserve and fast to be consumed (28). The use of this sub-employed protein source in meat product fabrication has an enormous potential since it is the main meat producer in rural areas of developing countries. An understanding of the functional properties of goat meat proteins is necessary to utilize these cheap carcasses, introducing new products, using nontraditional protein sources, and improving existing products (29). Refrigerated and frozen storage can affect structural and chemical properties of muscle foods, influencing quality attributes of meat and meat products (30). The objective of a study carried out in our laboratory was to compare the emulsifying properties of protein extracts obtained from goat and beef and to determined the effect of temperature and storage time this functional property.

Material and Methods. Goat meat was obtained from a municipal slaughterhouse and beef samples from a local butcher. Fat and connective tissue was removed and pieces of 100-150 g were cut perpendicularly to the meat fibers according to Wagner and Añon (31). Samples were vacuum packed in Cryovac bags at -700 mBars, and randomly allocated to different treatments. A 2x3x4 factorial design was used where the factores and levels were: animal species (goat and beef), storage temperature (4, -20 and -63ºC) and storage time (0, 1, 2 and 3 weeks). The response variables were: protein solubility, emulsion activity index (EAI), emulsifying capacity (EC) and sulfhydryl group content (SH). The results were analyzed for analysis of variance and Duncan multiple range test using a SAS package adapted to a personal computer (SAS Institute, Inc.).

Protein extraction was adapted from the reports by Li-Chan *et al.* (32). Protein content was determined by the Biuret reaction (33) adjusted to 10 mg of protein/ml. Protein solubility was obtained by centrifuging 10 ml of the protein solution at 25,000g during 15 minutes. Emulsion activity index (EAI) was determined according to Pearce and Kinsella (34) and Voutsinas *et al.* (35) using the correction reported by Cameron *et al.* (36). Emulsion capacity (EC) was determined by electric conductivity, based on the work of Swift *et al.* (37). Sulfhydryl group (SH) content was analyzed according to the method developed by Cofrades (38). Sunflower oil for human consumption was used in all determinations. A conventional domestic homogenizer was employed in the emulsion formation with an approximately speed of 12,000 rpm.

Results and Discussion

Analysis of variance (Table I) indicated that animal species had a significant effect on EAI (P>0.0001) and solubility (P>0.049) (Table I). This can be due to the time the meat was subjected to postmortem proteolysis, as goat meat were around 8 h postmortem whereas beef was 36 h. This resulted in more solubilized protein in goat (X=43.146) as compared to beef (X=38.279) (Table II). Endogenous enzymes in postmortem conditions could be more active in goat than in beef, although this fact needs to be clarified. Solubility affects also EAI, promoting higher values in goat meat (X=2141.88) than in beef (X=1726.75). These results agree with Chattoraj *et al.* (39) and Turgut (40) who compared the emulsifying properties of goat protein extracts against several species, finding than goat meat proteins had better emulsifying properties. Changes in EAI with storage time are due to protein degradation during storage, goat meat extracts presented higher values at the studied temperatures (Figure 1). Changes in protein size are related to the total accessible area, if a configuration of spherical objects is assumed. However, the actual total accessible area is almost two times larger for proteins than for perfect spheres of the same size due that proteins have rough surfaces (41). EAI is the superficial area originated by and emulsified mass, but is not and absolute measure of the created area (42). This is because each system has a given composition due to conditions in a particular experiment such as mechanical variations from one equipment to another, and energy applied in the emulsion formation.

Storage time had significant effects on EAI, EC and SH (P>0.0001, 0.0003 and 0.0006, respectively). As proteins undergo a degradation process and more fractions of smaller size are available, the probability of these fractions to interact with one another is higher therefore the emulsifying process is more feasible to occur. This could be also the cause of goat meat having higher EC values (X=73.0) tah beef (X=71.0) and was also related to protein hydrolysis in postmortem conditions. Nevertheless, peptides of very low molecular weight will be not adequate to form a stable emulsion. Storage time was the only source of variation giving a differences on the EC, conversely as reported by Jiménez Colmenero and Borderías (21) and by Kijowsky and Niewiarowic (22, 43) where EC decreased with time. Kijowsky and Niewiarowic (43) attributed this fact to differences in the film formation during the process of protein denaturalization. Values increased with storage time for all four response variables (Table III).

Temperature had no significant effect in any variable studied. EAI and EC did not give any information on the emulsion stability (44) due that drop size of the protein/oil system was not constant (35).

SH content increased with the storage time (P>0.0006). However, animal species did not give any significant difference with respect to this variable (Table III). Ancín *et al.* (45, 46) reported a high statistical correlation between the EAI and the SH groups in meat batters. Protein denaturation in meats, either by aging or processing, influence SH content. As it is related to protein conformation it also

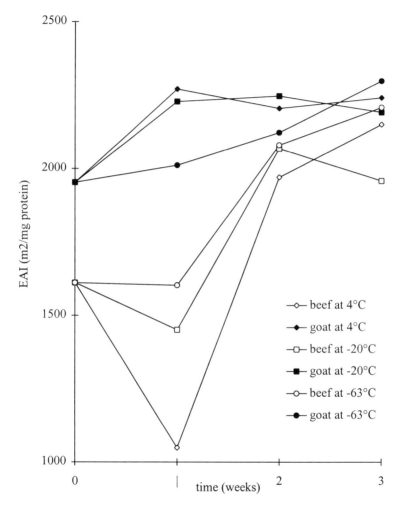

Figure 1. Emulsifying Activity Index (EAI) for goat and beef protein extracts

determines surface activity of proteins. This characteristic can be measured by EAI which is also related to structural characteristics of the meat proteins (46)

Table I. Emulsifying properties: Analysis of variance

Response variable	Source of variation (P>)		
	Animal species	Time	Temperature
EAI	0.0001	0.0001	0.7843
EC	0.1996	0.0003	0.5284
SH	0.3373	0.0006	0.7832
Solubility	0.0494	0.2546	0.1949

Conclusions

Emulsifying properties of goat meat extracts present values as acceptable as beef extracts. In this way, goat meat is a suitable alternative source of meat protein to fabricate emulsified meat products. The search for cheap proteins with suitable functional properties had focused the attention of many research workers to alternative sources, as sub-utilized species or slaughtering by-products.

Table II. Duncan Multiple Range Test: animal species

Response variable	Animal species		Standard error
	goat	beef	
EAI	2141.88[a]	1726.75[b]	44643.77
EC	73.6[a]	71.0[a]	47.7312
SH	39.405[a]	41.635[a]	69.3070
Solubility	43.146[a]	38.279[b]	63.3453

[a,b]: means with different superscript are significantly different (P>0.05)

The use of these protein materials is not only dictated by the need to increase supply of animal protein, but also by economical considerations. Protein materials other than myofibrillar, such as viscera and blood plasma and their processed forms are usually cheap (47). The development of new processes to fabricate protein concentrates and isolates must be carried out in order to provide protein products with improved functional and sensory attributes (48). Food additives of animal origin depends on substituting myofibrillar proteins, in case of substitution technology, or to be used as an ingredient in a product formulation. However, reports on protein functionality are not reproducible among different laboratories due to lack of homogeneity in the methodology employed.

Table III. Duncan Multiple Range Test: time

Response variable	Weeks				Standard error
	0	1	2	3	
EAI	1370.72[b]	1769.74[b]	2118.04[a]	2178.76[a]	49643.7
EC	64.975[c]	76.283[ab]	70.958[b]	76.983[a]	47.7312
SH	37.748[b]	36.063[b]	38.723[b]	49.548[a]	69.3070
Solubility	37.675[a]	39.642[a]	41.100[a]	44.433[a]	63.3253

[a, b, c]: means with different superscript are significantly different ($P > 0.05$)

Acknowledgments

Author Totosaus thanks the National Council for Science and Technology (CONACYT) Mexico, for a Master's degree scholarship.

Literature Cited

1. Pour-El, A. In *Protein Functionality in Foods*; Cherry, J.P., Ed.; ACS Symposium Series 147; American Chemical Society: Washington, D.C., 1981: 1-12.
2. Wilding, P.; Lillford, P.J.; Regenstein, J.M. *J. Chem. Tech. Biotechnol.* **1984,** 34B.
3. Whiting, R.C. *Food Technol.* **1988,** 42, 104.
4. Kinsella, J.E. In *Food Proteins*; Fox, P.F.; Condon, J.J., Eds.; Applied Science Publishers: London, United Kingdom, 1982; 51-65.
5. Smith, D.M. *Food Technol.* **1988,** 42, 116.
6. Borderías, A.J; Montero, P. *Rev. Agroquím. Tecnol. Alim.* **1988,** 28, 159.
7. Damodaran, S. In *Protein Functionality in Food Systems.*; Hettiarachchy, N.S.; Ziegeler, G.R., Eds.; Marcel Dekker Incorporated: New York, 1994; 1-37.
8. Rodriguez, F.J. *Ind. Alimentaria* **1985,** 7, 4.
9. Kretzschmar, U. *Fleischwirstch.* **1992,** 72, 905.
10. Lawrie, R.A. *Ciencia de la Carne.* Editorial Acribia, Zaragoza, 1974, 91-98.
11. Ranken, M.D. In *Developments of Food Protein:Volume 3*; B.J.F. Hudson, Ed.; Elsevier Applied Science Publishers: London, 1985; 1-35.

226

12. Tarrant, P.V. In *Foods Proteins*;. Fox, P.F.; Condon, J.J., Eds.; Elsevier Applied Science Publishers: London, 1982; 261-291.

13. Woloszyn, J. *Godosporka-miesa* **1992**, 44, 22.

14. Morrissey, P.A.; Mulvihill, D.M.; O'Neill, E. M. In *Developments in Food Proteins: Volume 5*, Hudson, B.J.F., Ed.; Elsevier Applied Science Publishers: London, 1987; 195-256.

15. Xiong, Y.L. *Crit. Rev. Food Sci. Nut.* **1994**, 34, 293.

16. Lan, Y.H.; Novakofski, J.; Carr, T.R.; Mckeith, F.K. *J. Food Sci.* **1993**, 58, 963.

17. Cheftel, J.C.; Cuq J.L.; Lorient, D *Protéines Alimentaries*. Technique et Documentation- Lavoiser: Paris, 1985; 45-65.

18. Saffle, R.L. *Adv. Food Res.* **1968**, 16, 105.

19. Lee, C.M.; Carroll, R.J.; Abdollahi, A. *J. Food Sci.* **1981**, 46, 1789.

20. Schmidt, R.H. In *Protein Functionality in Foods*; Cherry, J.P., Ed.; ACS Symposium Series 147; American Chemical Society: Washington, D.C., 1981; 131-147.

21. Jiménez-Colmenero, F.; Borderías, A.J. *J. Food Technol.* **1983**, 18, 731.

22. Kijowski, J.; Niewiarowicz, A. *J. Food Technol.* **1978**, 13, 451.

23. Chen, P.H.; Ockerman, H.W. *Meat Focus Intl.* **1995**, 4, 235.

24. López, C.; Obaya A.; Meléndez, R. *Tecnol. Alim.* **1995**, 30, 11.

25. Gillet, T.A.; Mieburg, D.E.; Brown; C.L.; Simon, S. *J. Food Sci.* **1977**, 42, 1606.

26. Sulzbacher, W.L. *J. Sci. Food Agric.* **1973**, 24, 589.

27. Einsminger, M.E.; Parker, R.O. Sheep and Goat Science; The Interstate Printers and Publishers Inc.: Danville, Illinois, 1986; 291-296.

28. Arbiza, G.E. Producción de Caprinos, AGT Editor, S.A.: México, 1986; 129-150.

29. Smith, D.M. *Food Technol.* **1988**, 42, 116.

30. Miller, A.J.; Ackerman, A.; Palumbo, S.A. *J. Food Sci.* **1980**, 45, 1466.

31. Wagner, J.R.; Añon, M.C. *J. Food Technol.* **1985**, 20, 735.

32. Li-Chan, E.; Nakai, S.; Wood, D.F. *J. Food Sci.* **1985**, 50, 1034.

33. Gornal, A.G.; Bardawill, C.J.; Davis, M.M. *J. Biol. Chem.* **1949**, 177, 751.

34. Pearce, K.N.; Kinsella, J.E. *J. Agric. Food Chem.* **1978**, 26, 716.

35. Voutsinas, L.P.; Cheung, E.; Nakai, S. *J. Food Sci.* **1983**, 48, 26.

36. Cameron, D.R.; Weber, M.E.; Idziak, E.S.; Neufeld, R.J.; Cooper, D.G. *J. Agric. Food Chem.* **1991**, 39, 655.

37. Swift, C.E.; Lockett, C.; Fryar, A.J. *Food Technol.* **1961**, 15, 468.

38. Cofrades, S. Tesis Doctoral, Universidad Complutense, Madrid, Spain, 1994.

39. Chattoraj, D.K.; Bose, A.N.; Sen, M.; Chatterjee, P. *J. Food Sci.* **1979**, 44, 1659.

40. Turgut, H. *J. Food Sci.* **1984**, 49, 168.

41. Creighton, T.E. *Proteins*; W.H. Freeman and Company: New York, 1984; 239-245.

42. Das, K.P.; Kinsella, J.E. *Adv. Food Nut. Res.* **1990**, 34, 81.

43. Kijowski, J.; Niewiarowicz, A. *J. Food Technol.* **1978,** 13, 461.
44 Arteaga, G.E. Ph. D. Thesis, The University of British Columbia, Columbia, Canada, 1994.
45. Ancín, C; Sánchez-Monge, J.M.; Villanueva, R.; Bello, *Anal Bromatol.* **1989,** 41.
46. Ancín, C; Sánchez-Monge, J.M.; Villanueva, R.; Bello, *Anal Bromatol.* **1989,** 41.
47. Duda, Z. *Przemysl spozywczy* **1993**, 5, 135.
48. Zayas, J.F. *Functionality of Food Proteins.* Springer: Berlin, 1997;1.

Fat and Oil Functionality: Physiological Functionality

Chapter 15

Physicochemical Aspects of Triacylglycerides and Their Association to Functional Properties of Vegetable Oils

Jorge F. Toro-Vazquez and Miriam Charó-Alonso

Facultad de Ciencias Químicas-CIEP, Universidad Autónoma de San Luis Potosí, Av. Dr. Manuel Nava 6, Zona Universitaria, San Luis Potosí, México, 78210

The molecular phenomena that determine the viscosity and melting/crystallization temperatures of triacylglycerides (TG) in vegetable oils are discussed. These physico-chemical properties determine some of the functional properties of vegetable oils when used alone or in complex food system. For instance, when oils are cooled, the TG family with the highest melting temperature is the first to crystallize developing a solid in a liquid phase. These systems, generally known as plastic fats, have fractal organization with significant effect on the melting properties and spreadability of food systems like butter and margarine. On the other hand, in vegetable oils TG's are organized in bi-layer lamellar structures whose shape and size change with temperature and determine the magnitude of oil viscosity. In turn TG crystallization is affected by viscosity which determines, in a great extent, nucleation rate and crystal growth.

Lipids are a heterogenous group of compounds mainly composed of carbon, hydrogen, and oxygen, and are soluble in non-polar solvents and insoluble in water. However, the term lipid has not a strict definition. Christie (*1*) indicated that: "Lipids are fatty acids and their derivates and substances related biosynthetically or functionally to these compounds." This definition will be adopted throughout this chapter.

From the physiological point of view lipids have basic functions such as supplying energy (9 Kcal/g, value traditionally assigned and derived from caloric measurements) or storing it for further utilization, providing essential fatty acids for human metabolism, and providing a transport medium for liposoluble vitamins. On the other hand, the particular functional properties of lipids provide foods with specific characteristics of texture, flavor, odor, and color. These functional properties result from a lipid's distinctive physical chemical characteristics, and from their interactions with other food components as affected by process variables. Such is the case of the interaction of lipids with starch (*2-5*), milk and meat proteins (*6-10*), water (*11, 12*), or even other lipid molecules (*13-15*).

Considering the physical and chemical behavior of lipids, it is common to divide them into polar and non-polar lipids. However, none of the lipids included in each category is completely polar or non-polar, and the classification just indicates the predominant molecular property (i.e., lipophilic or hydrophilic) that determines its degree of solubility in polar or non-polar solvents. Thus, polar lipids (i.e., phospholipids, galactolipids, and monoglycerides), that have a polar element (e.g., a phosphate in the 3 sn-position of glycerol in the case of phospholipids), interact quite easily with water developing aqueous phases in lipidic environments. This amphiphilic nature of polar lipids is also reflected in the way the molecules organize in the crystalline form and liquid phases (i.e., $L2$ phase), which results in structures with application in cosmetics, drug delivery products (*11, 16*), and food coating (*17, 18*). In contrast, non-polar lipids do not form lipid-water phases since their structure is mainly a hydrocarbon backbone. The group of non-polar lipids includes the major component of vegetable oils and animal fats, the triacylglycerides (TG). The information presented here deal with some of the physical chemical characteristics of this particular group of lipids and their association with functional properties of vegetable oils.

Triacylglycerides are triesters of glycerol with fatty acids, and were originally named triglycerides (Figure 1). However the use of this term is now discouraged (*19*). When glycerol is esterified in one or two of the positions the polarity of the molecule increases, and the compound is known as mono- and di-acylglycerides respectively (Figure 1). In consequence, mono- and di-acylglycerides are considered polar lipids. TG's have several physical chemical properties that determine processing conditions of fats and oil, as well as the organoleptic and stability properties of the food they constitute. The phase changes of TG, i.e., polymorphic transformations and melting and crystallization temperatures, are physical properties that determine the functionality that lipids provide to food systems such as texture, spreadability, coating, etc. In general, physical properties are associated with chemical parameters that describe the shape and organization of the molecules in the system. In TG such chemical parameters are, among the most important, the degree of unsaturation and length of the hydrocarbon chain of the fatty acids, as well as the positional distributions of the fatty acids esterified to the glycerol (i.e., stereospecific distribution) (*11*). In this chapter the viscosity and phase changes that occur in vegetable oils are discussed, associating the behavior of such physical properties to chemical characteristics of TG and their interaction with the time-temperature variable to determine particular functional properties of vegetable oils.

Vegetable Oil Viscosity, General Concepts

Viscosity is a measure of the resistance of a liquid to flow. The flow properties (i.e., rheology) of food components determine, in a great extent, the texture and stability of food systems. In the same way, rheological parameters are required in the design of several operations utilized during food processing. For example, oil viscosity is utilized in the estimation of distillation column efficiency for separation of fatty acids, mono-, and di-acylglycerides, during the control of hydrogenation and interesterification, as well as designing the piping for oil transportation (*20-22*). Additionally, oil viscosity is also an important parameter that determines the rate of diffusion of compounds such as free fatty acids and carotenoids during their separation through the adsorption process of bleaching in vegetable oil refining *(23, 24)*. In the same way, diffusion of TG during crystallization of vegetable oils is influenced by viscosity of the liquid phase, affecting the crystal growth rate and size, induction time for crystallization, and melting/crystallization temperature of the crystals developed *(25-27)*. Thus, viscosity determines the efficiency and rate of the reactions involved in process such as vegetable oil bleaching, winterization, and fractional crystallization.

In general and from a molecular point of view, viscosity is a macroscopic measurement of the internal friction between the molecules that constitute the fluid and which oppose to movement. This interaction among the molecules changes with temperature, and in consequence, viscosity is a function of temperature, generally decreasing exponentially with the increase in the temperature of the system. Therefore, viscosity has significant correlations with structural parameters of the fluid molecules and such relationships must change as a function of temperature. However, with lipids and in particular with vegetable oils, few studies have addressed this issue *(21, 28, 29)*.

Effect of Temperature.   Vegetable oils follow Newtonian flow behavior, i.e., at a given temperature (T°) a constant value of viscosity is obtained independent of the force applied to the oil. As a result, fluid or oil viscosity (η) is defined as the slope of the shear stress-shear rate curve, also known as the flow curve (i.e., force applied vs. flow rate achieved by the oil). Flow curves for sunflower oil and a mixture of sunflower oil with a margarine are shown in Figure 2.

Controlled shear rate viscometers are utilized to obtain flow curves, and the most commonly utilized is the Brookfield viscometer. However, since oil viscosity is independent of the shear rate applied (i.e., Newtonian fluid) capillary viscometers such as the Ubbelohde or Cannon Fenske have been also utilized *(27, 30)*. When the driven force is gravity, like in capillary viscometers, fluid or oil viscosity may be obtained as kinematic viscosity, ν, making oil density (ρ) an additional parameter to consider. In consequence at a given temperature ν and η are associated through equation 1.

$$\nu = \eta/\rho \qquad (1)$$

In vegetable oils, η and in turn ν, decreases with T°. However the extent and pattern of this decrease depend on the composition of the oil *(29)*. An Arrhenius type equation has been used to describe the effect of T° on oil viscosity. However, the predictions obtained have been unsatisfactory when compared to measured data. Additional models utilized to evaluate the T° effect on oil viscosity from different vegetable sources include modifications of the Arrhenius equation, like the Andrade *(27, 29, 30)* and the Antoine model *(21, 27, 29)*. Table I shows the values of the constants A, B, and C for several vegetable oils obtained utilizing the Antoine's equation (equation 2) and different extensions of the Andrade's equation (equations 3 and 4). The predicted efficiency of these equations is restricted to the given vegetable oil and interval of T° to which the particular equation was developed. The different magnitudes of the A, B, C constants obtained in diverse studies (Table I) indicate the sensitivity of such constants to

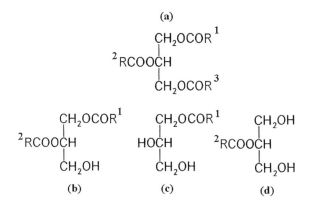

Figure 1. Representation of a triacylglycerol molecule (a), also called triglyceride or neutral lipid. 1, 2,- diacylglycerol (b), 1,- monoacylglycerol (c), and 2,- monoacylglycerol are also represented.

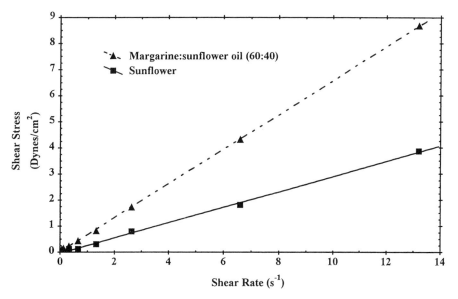

Figure 2. Flow curves for sunflower oil and a mixture (60:40) of a margarine with sunflower oil. The curves were determined at 298 K in a Brookfield viscometer (adapted from ref. 29).

Table I. Correlations Constants for Prediction of Dynamic (η) or Kinematic (v) Viscosity for Several Vegetable Oils.

Vegetable oil	Constants			Temperature interval (°C)	Viscosity[a] (cP)	Equation (reference)
	A	B	C			
Corn	-19.5604	5225.44	0.019943	23.9 - 110	v	4 (30)
Soybean	-20.0596	5323.74	0.020634	23.9 - 110	v	4 (30)
Coconut	1.0945	-2251.54	917780	37.8 - 110	v	3 (30)
Rapeseed	-24.4621	6210.21	0.026671	23.9 - 110	v	4 (30)
Canola	-4.4068	2501.109	5098.307	25.0 - 65	η	3 (29)
Sesame	-4.2709	2485.478	-4907.076	25.0 - 65	η	3 (29)
Corn	-4.9536	2677.526	-311.844	25.0 - 65	η	3 (29)
Canola	6.2246	5060.580	1319000	0.0 - 40	v	3 (27)
Several[b]	-0.6298	273.66	88.81	0.0 - 100	η	2 (21)

[a] Viscosity measurement in centi-Poises (cP).
[b] A generalized model including 26 different oils.

the measurement conditions and model utilized. Additionally, these constants are just fitting parameters that do not have a particular physical chemical meaning associated to the T° effect on oil viscosity.

$$Ln(\eta) \text{ or } Ln(\nu) = A + B/(T^\circ + C) \tag{2}$$

$$Ln(\eta) \text{ or } Ln(\nu) = A + B/T^\circ + C/(T^\circ)^2 \tag{3}$$

$$Ln(\eta) \text{ or } Ln(\nu) = A + B/T^\circ + C(T^\circ) \tag{4}$$

Effect of Unsaturation and Chain-length of Fatty Acids. Dutt and Prasad (*21*) analyzing published oil viscosity data from 21 different sources observed that the unsaturation degree of the esterified fatty acids, value directly associated with the iodine value (IV), has an inverse relationship with the viscosity of the oil. In the same way the authors reported that the chain-length of the fatty acids that constitute the triacylglycerides and indirectly correlated with the saponification value (SV), has a direct relationship with oil viscosity. Under this basis the authors developed a generalized equation (equation 5) with three parameters: T°, IV, and SV. The development of the equation was based on the IV/SV ratio, since each parameter has an opposing effect on viscosity. However, this assumption did not consider the independent effect of fatty acid unsaturation (e.g., IV) and chainlenght (e.g., SV), nor their interactions as a function of T° on vegetable oil viscosity. This particular point was later addressed by Toro-Vazquez and Infante-Guerrero (*29*), who utilizing a multiple variable regressional approach for 21 vegetable oils developed a generalized model applicable for the 25°C - 65°C interval (equation 6). Based on this model particular conclusions were made that relate the TG structure to the magnitude of vegetable oil viscosity. Thus, as the concentration of the naturally occurring *cis* double bond increases in the oil (i.e., increasing the IV), the intermolecular interactions among the TG's in the liquid state are hindered, since this double bond configuration produces a kink in the natural "linear" zig-zag organization of the hydrocarbon chain in the saturated fatty acid molecules in the TG. In consequence, both the oil viscosity and the T° of solidification of the oil decrease as IV increases (*29*). According to equation 6 the *cis* double bond effect (i.e., unsaturation degree of vegetable oils) on oil viscosity is independent of both the chain-length of the fatty acids (i.e., SV) and the T° of the system (Figure 3). In contrast, the effect of the length of the esterified fatty acids on oil viscosity is, according to equation 6, a function of the T° of the system (Figure 4). Thus, as the T° decreases in the interval within 25°C to 45°C, the families of TG constituted by the shortest fatty acids (i.e., higher SV) are prone to organize in liquid lamellar structures (*31*), previous to their crystallization, increasing oil viscosity as temperature decreases (Figure 5) (*11, 25, 29*). At the higher T° interval (> 45°C) the size of such triacylglyceride lamellar structures decreases (Figure 5) making the chain-length effect on oil viscosity less significant (Figure 4) (*29*). The importance of these phenomena during TG crystallization has been studied through reduced viscosity measurements in oil solutions of saturated TG (i.e., tripalmitin and tristearin) in sesame oil (*25*).

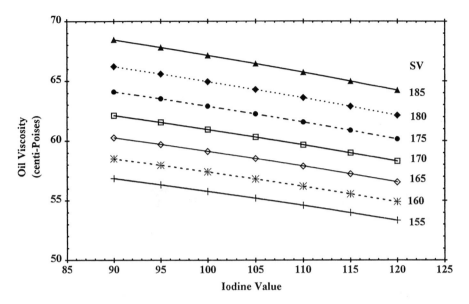

Figure 3. Oil viscosity as a function of Iodine value and saponification value. The temperature was 298 K and the viscosity was calculated with equation 6 (adapted from ref. 29).

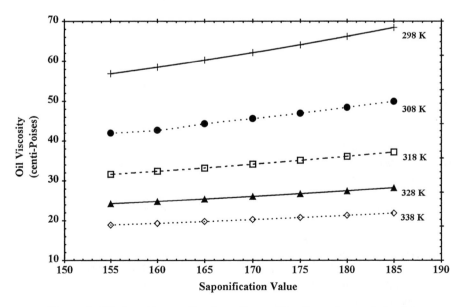

Figure 4. Oil viscosity as a function of saponification value and absolute temperature. The Iodine value was kept constant to a value of 90 to calculate the viscosity with equation 6 (adapted from ref. 29).

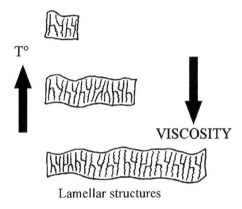

Figure 5. *Lamellar structures of triacylglycerides in the liquid state as a function of temperature.*

Further studies with equation 6 to describe the viscosity of vegetable oils to temperatures down to -10°C resulted in equation 7. In this equation, C_{16} is the concentration (% w/w) of palmitic acid and $C_{18:3}$ is the concentration of linolenic acid (% w/w). Equation 7 is applicable in a wider temperature interval going from -10°C up to 65°C, and indicates that at T°'s lower than 25°C the concentrations of palmitic and linolenic acid are the main variables that determine vegetable oil viscosity. Thus, within this T° interval the oil viscosity increases in a direct relationship with C_{16} as temperature decreases quadratically [i.e., variable $(C_{16})/(T°)^2$ in equation 7]. In contrast, viscosity decreases in an indirect relationship with $C_{18:3}$ as temperature increases [i.e, variable $T°/(C_{18:3})$ in equation 7]. At temperatures above 25°C there is not a significant effect of a particular fatty acid and equation 6 is applicable.

$$Ln(\eta) = [-1.4 + 1.25(IV/SV)] + [500 - 375(IV/SV)]/[(T° + 140) - 85(IV/SV)] \quad (5)$$

$$Ln(\eta) = -4.7965 + 2525.92962(1/T°) + 1.6144(SV)^2/(T°)^2 - 101.06 \times 10^{-7}(IV)^2 \quad (6)$$

$$Ln(\eta) = -4.7965 + 2525.92962(1/T°) + 1.6144(SV)^2/(T°)^2 - 101.06 \times 10^{-7}(IV)^2 + 2685.0197(C_{16})/(T°)^2 + T°/(C_{18:3}) \quad (7)$$

Reduced Viscosity Measurements in Vegetable Oils. The reduced viscosity measurement (η_{red}) is a parameter associated with changes in the molecular conformation of solute molecules (i.e., intramolecular phenomena) as a function of its concentration in a solvent, as well as with changes in the intermolecular interactions (e.g., association, dissociation, etc.) between solute molecules and between solute and solvent molecules (32). For Newtonian liquids this parameter is just a function of T° and it is calculated with equation 8, where η_o is the viscosity of the pure solvent, η_{Sol} is the viscosity of a solution with a solute at concentration C. In the particular case of TG's their changes in molecular conformation in solution are insignificant and therefore their contribution to the magnitude of η_{red} is negligible. Nevertheless, the effect of the addition of a particular TG (i.e. solute) on the original triacylglyceride lamellar organization in a given vegetable oil (i.e., solvent) has practical implication on the magnitude of oil viscosity and TG crystallization, which might be investigated through η_{red} measurements. In fact such measurements have been performed in tripalmitin and tristearin sesame oil solutions (25). The results have indicated that the increase in interfacial energy produced when the saturated TG is segregated out of the unsaturated lamellar triacylglyceride organization of sesame oil as temperature decrease, is described by the change in η_{red} (25). As a result the temperature of crystallization of the TG of highest melting temperature in the oil system (i.e., solute) is achieved when $\eta_{red} = 0$ (Figure 6), which represents the point where the interfacial energy between the developing crystal nucleus and the unsaturated TG of sesame oil (i.e., solvent) achieves its maximum value (i.e., minimum solubility of the crystal nucleus). Thus, the higher the intermolecular interactions between the TG of highest melting temperature and the rest of the TG in the oil system (i.e., similar or complementarity conformations between different families of TG's), the lower the T° needed to accomplish the point where $\eta_{red} = 0$. In consequence, the onset of crystallization occurs at a lower T° making the viscosity of the liquid phase a limiting factor for the growth of the crystal nucleus,

A

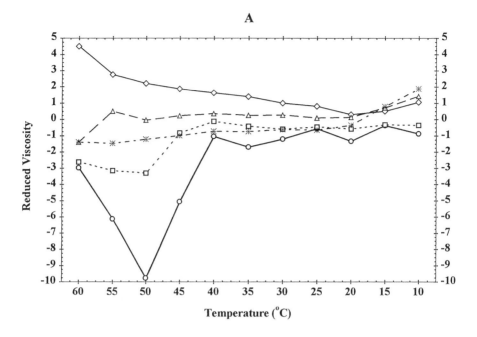

B

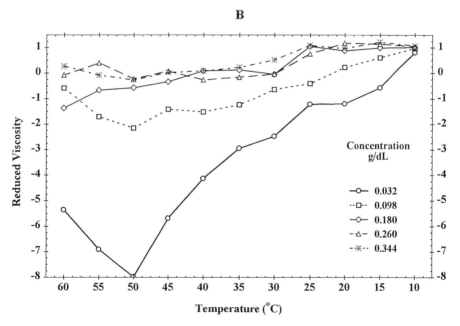

Figure 6. Reduced viscosity for tripalmitin (A) and stearin (B) solutions as a function of their concentration in sesame oil and temperature. The shear rate utilized for the determination was 7.92 s⁻¹ (adapted from ref. 25).

since diffusion of the crystallizing molecule is an inverse function of viscosity (*33*).
Negative values of η_{red} are sometimes obtained since the addition of a particular TG to an
oil may disrupt the original lamellar triacylglyceride structure, producing shorter lamellar
structures and thus η_{Sol} is smaller than η_o (equation 8).

$$\eta_{red} = [(\eta_{Sol} - \eta_o)/\eta_o](1/C) \qquad (8)$$

Concentrated miscella oil bleaching is another process in which viscosity and its
association to molecular diffusion play an important role. In this refining process the
concentration of oil in the extraction solvent is 40 to 80% (*34*). At these concentrations
diffusional factors (e.g., oil viscosity) limit the achievement of thermodynamic adsorption
equilibrium, a condition that has always been assumed valid for the application of the
adsorption models utilized by the oil industry (i.e., the Freundlich and Langmuir
isotherms) (*23, 24*). In adsorption studies performed in a concentrated miscella system
with sesame oil (60 to 100% oil) using vegetable carbon as adsorbent, higher adsorption
of free fatty acids and carotenoids were achieved as solve4nt concentration was increased
(*23, 24*). The viscosity of the miscella followed an inverse relationship with solvent
concentration according to equation 9 (*24*). Thus, as viscosity decreases in the oil miscella
system the diffusion of adsorbates toward the surface and through the pores of the
adsorbent is promoted, increasing the access of the adsorbate (i.e., free fatty acids and
carotenoids) to the adsorbing sites in the adsorbent.

$$1/\eta = 10.4014 \times 10^{-3} + (4.283 \times 10^{-3})(\% \text{ solvent}) \qquad (9)$$

Phase Changes of Triacylglycerides in Vegetable Oils.

In general, the phase changes of TG's are physical properties that determine the solid-to-
liquid ratio provided by TG in the oil as a function of $T°$. Since these properties determine
much of the functionality of edible vegetable oils in a given food system, their relationship
with the chemical structure of TG's, and the functional characteristic of interest in food
systems must be established. However, vegetable oils are mixtures of different families
of the TG, and consequently do not have specific physical properties. For instance,
vegetable oils do not show a particular melting or crystallization temperature. Rather,
vegetable oils show a melting/crystallization $T°$ profile. Thus, when oils are cooled the TG
family with the highest melting $T°$ (i.e., the more saturated TG) is the first to crystallize
developing a solid in a liquid phase. These lipid systems, generally known as plastic fats,
have fractal organization, i.e., a TG crystal network in an oil continuum (*35*). Additionally, TG's crystallize in different polymorph states (i.e. different TG molecular
organization corresponding to different packing of the hydrocarbon chain), which have
distinctive degrees of thermal stability, crystal size, and shape. These polymorph states for
pure TG crystals basically are, in increasing order of thermal stability, sub-α, α, β', and
β. Polymorphic forms differ in the geometry of packing of the hydrocarbon chain at the
molecular level. At a higher molecular level the TG arrangement in the solid state may
occur, according to Larsson (*11, 12*), in four different types (Figure 7). Two of these

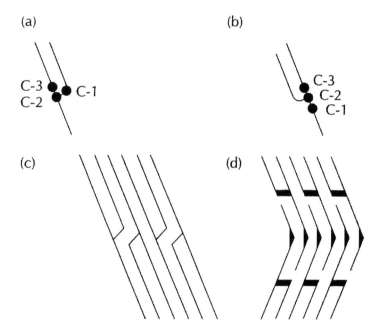

Figure 7. Triacylglycerol molecular arrangements in the solid state. The circles represent the three glycerol carbon atoms and the lines the fatty acyl hydrocarbon side chain. The bend in the triacylglyceride arrangement indicated in (d), results from the presence of a double bond in one of the fatty acid chains (adapted from ref. 11, see also ref. 12).

structures depend on the conformation of the fatty acyl groups on the glycerol backbone (structures A and B in Figure 7), the other two on the organization of hydrocarbon chains of the TG's across the different layers of the crystal (structures C and D in Figure 7). The A and B type molecular arrangements can be organized into either C or D type organizations, yielding the four possible types of TG structures (*11, 12*).

Within this framework, the phase changes of TG's are melting and crystallization temperatures, as well as polymorphic transformations. In consequence, the functional properties associated with the use of vegetable oils (i.e., mouthfeel, spreadability, emulsion stability, whippability, and air bubble retention capacities) in products such as chocolate, butter, low fat spreads, shortenings, ice cream and whipped cream, depend on the capacity of the TG's in the oil to develop a solid phase, its melting/crystallization T° profile, polymorph state, and fractal organization of the crystallized TG in the oil as affected by the time-temperature conditions. The effect of time-temperature conditions on vegetable oils functionality is emphasized since TG organization in the solid state is a metastable process (*11, 26*). Thus, a slight variation in the time-temperature conditions has a profound effect on the kinetics of polymorphic transformations, and subsequently in the amount of solid phase in the system, and its melting/crystallization T° profile.

Determination of Phase Changes in Vegetable Oils.   The solid-to-liquid ratio in a vegetable oil is evaluated through the solid fat index (SFI), parameter extensively utilized by the industry which actually quantifies the solid to liquid TG ratio in the oil as a function of temperature. SFI curves are determined by dilatometry, differential scanning calorimetry (DSC), and nuclear magnetic resonance. The last two methods are nowadays widely utilized. Nevertheless, nuclear magnetic resonance is a more accurate technique. The DSC analysis assumes, erroneously, a constant heat of fusion of all TG's in the oil system (*36*-38). The heat of fusion of TG's increases almost linearly with the melting T° of TG fractions, and thus the real amount of crystallization will differ depending on the T° interval utilized for the calculation of the SFI (*36*). In contrast, the melting/crystallization T° profile is most of the time evaluated through DSC and different time-temperature conditions have been utilized to perform this analysis (*26, 36*). A collaborative study was organized by the Physical Methods Committee of the American Oil Chemist's Society to standardized the time-temperature conditions and the DSC calibration procedure. The objective was to decrease the variability in the determination of melting peak temperature, onset temperature for crystallization, entalphy of fusion (ΔH_F), temperature of completion of melt, and evaluation of polymorphic behavior. The result of this study was the establishment of the recommended practice number Cj 1-94 for the determination of the melting/crystallization temperature profile through DSC (*39*). Nevertheless, further collaborative studies are needed before an AOCS official method is adopted. Figure 8 shows the melting and cooling DSC thermogram for tripalmitin obtained utilizing the methodology recommended by the AOCS (*39*). The polymorphic transformations that occur during tripalmitin melting, which point out the metastable character of TG crystallization, are indicated in this Figure. On the other hand, during the last years it has become evident that in addition to DSC, X-ray diffraction is a powerful analytical technique to identify polymorphic phases unambiguously in both pure TG systems and vegetable oils (*40-42*). Although the original technique requires long exposure times of the sample, recent developments in high-energy accelerators (i.e., synchrotrons) and X-ray detectors have reduced the time to the order of milliseconds, and soon enough to microseconds (42). With synchrotron radiation X-ray diffraction, the kinetics of rapid TG polymorphic transformations has been elucidated under both isothermal and nonisothermal conditions in pure and mixed TG systems (43-45).

The fractal nature of TG crystals might be evaluated through the measurement of the elastic modulus (G') of the crystallized system, provided the crystal volume fraction is known. An alternative methodology involves the use of light scattering (35). In any case, the complex structure of the crystal network is determined by the fractal dimension, D, which describes the relation between the number of crystals in a crystal aggregate (e.g., flock) and its radius, R, i.e., $N \sim R^D$ (35, 46). In general, the higher the value of D, the more compact the crystal dispersion and, in the case of TG, its value changes with ageing of the system after crystallization. In turn, the magnitude of D affects the value of G' (35) and therefore the texture of the crystal dispersion at a given temperature (e.g., mouthfeel, spreadability). Although the rehology of fat crystal dispersions determines important quality properties in vegetable oil system (i.e., texture, sedimentation), still more research is needed to understand the structure-property relationship. Within this context the application of fractal models, although still in early development, seems promising.

Importance of Crystallization in Vegetable Oil Processing. Three processes are utilized to modify the phase change properties of vegetable oils and improve the versatility of their utilization by the food industry. These process are hydrogenation, crystallization, and interesterification. However, consumer concern associated to the atherogenic effect of *trans* fatty acids (47), which are produced during the hydrogenation of the naturally occurring *cis* unsaturated fatty acids (i.e., oleic, linoleic, and linolenic fatty acids), limits the future of this process as a modification technique of the SFI in vegetable oils. Nevertheless, important developments directed to control the production of *trans* fatty acids during hydrogenation have been made recently (48, Machado, R.M.; Gaumer Freidl, K.; Achenbach, M.L. *J. Am. Oil Chem. Soc.*, submitted for publication, 1997). On the other hand, the chemical or enzymatic interesterification process are technologies utilized in European countries, nevertheless particular drawbacks mainly associated to the yield efficiency and cost of the processes have limited their use in the United States (49). In contrast, crystallization has been broadly utilized to produce value-added TG fractions from different fats and oils (50, 51); however, the economy efficiency of this modification process depends on the industrial application and market price of the secondary products obtained from such process (i.e., secondary fractions) (52). A slight variant of fractional crystallization frequently used in the refining oil industry is winterization; through this fractional process a small quantity containing high melting compounds is eliminated from some vegetable oils (27, 53), so they remain clear at low ambient temperatures increasing consumer acceptance.

During the last years fractional crystallization technology has been driven by the growth in the production of palm oil, which is further fractionated for the production of palm olein and palm stearin (52, 54, 55). Palm olein is mainly utilized as a frying oil and palm stearin is used in the manufacture of *trans* free margarine and vegetable shortening (54, 56). In the near future, palm oil will be the most economical and abundant edible oil worldwide (55), which provides the marketing incentive to increase our knowledge in the events involved in TG crystallization to further improve fractional crystallization technology. Additionally, there is an obvious need to develop food systems that eventually support the expansion in the utilization of palm oil or its fractions; crystallization of palm oil or palm stearin in mixture with other vegetable oils to produce spreads or margarine might be a feasible alternative (15, 25, 26). Tripalmitin is the triacylglyceride with the highest melting temperature in both palm oil (5-10% w/w) and palm stearin (12-56% w/w, depending on the fractionation temperature). In consequence, tripalmitin has a significant effect on crystallization kinetics and polymorphic behavior of palm oil and palm stearin.

Triacylglyceride Crystallization in Vegetable Oils. The fundamental concepts of the process of crystallization are beyond the objective of this chapter. These concepts are very well addressed by Boistelle (57) and Larsson (11).

It is generally accepted that three different events are involved during crystallization, namely nucleation (solid phase formation), crystallization (crystal growth), and crystal ripening (crystal perfection). These events happen simultaneously or nearly so at different rates, since there is a continuous variation of the conditions that produce crystallization (57). However, before nucleation occurs, the solution must be under supercooling or supersaturation conditions in order to thermodynamically drive the formation of a solid nucleus. Nevertheless, there is not a clear distinction between the effect of supercooling and supersaturation on crystallization, specially when the conditions are utilized at industrial scale (58). Supercooling, ΔT, is defined by $\Delta T = (T^{\circ}_M - T^{\circ}_C)$, where T°_M is the melting temperature of the crystallizing compound and T°_C is the isothermal crystallization temperature of the system. In the case of pure triacylglycerides T°_M is measured by DSC as its melting peak temperature (Figure 8), and with mixed TG's and vegetable oils as the peak temperature of the component of highest melting temperature. In contrast, several quantities are commonly utilized to measure the supersaturation extent, these are β, Ln β, $(\beta - 1)$ and C - C_s), where $\beta = C/C_s$ and C is the concentration of the compound in the solution at a given T° (e.g., the crystallization temperature), concentration higher than the concentration at saturation C_s at the same temperature. For crystallization to occur T°_C must be lower than T°_M, and C must be higher than C_s at the crystallization temperature. Additionally, as supercooling or supersaturation increases the induction time for crystallization decreases while nucleation rate increases (25, 33, 57, 58). This relationship holds until the decrease in T°_C or the increase in C is such that viscosity of the liquid phase limits diffusion of the crystallizing molecules, making the induction time for crystallization going up to impractical values.

The concept of supercooling is normally utilized when crystallization is achieved from a melt of one (i.e., pure TG), or often two or more components (i.e., vegetable oil), and the objective is either a simple solidification of a one-component system but most of the times the purification of a multicomponent system (e.g., mixed crystals). In contrast, the term supersaturation is utilized when crystallization of a particular substance (i.e., a given family of TG) is achieved from a solution (i.e., vegetable oil in hexane or acetone), and its objective is the separation of a pure compound. In any case, supercooling and supersaturation are associated with the developing of thermodynamic conditions needed

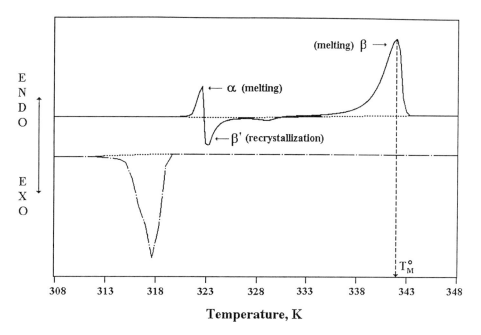

Figure 8. Melting (solid line, heating rate 5 K/min) and cooling profiles (dotted line, cooling rate 10 K/min) for pure tripalmitin obtained by differential scanning calorimetry. The polymorph transformations during melting of tripalmitin are indicated as α, β', and β.

to structure the molecules in a liquid phase ("liquid structure," 25), until a critical size of monomers (i.e., TG) aggregates and solid nucleus are formed (11, 57). In the liquid state, TG's are organized in bilayer lamellar structures (11) (Figure 5), the shape and size of these structures determine the change of viscosity with the T° of the system. Thus, oil viscosity is an indirect measurement of the degree of organization achieved by TG's in the liquid state. The relation between the structural organization of molecules in the melt or in solution and the crystallization process have been recognized for a long time. This aspect was studied through η_{red} measurements (equation 8) in nonisothermal crystallization of tripalmitin and tristearin in sesame oil (Figure 6) (25), as has been previously mentioned. In general, crystallization of the TG of highest melting temperature in the oil system is achieved when $\eta_{red} = 0$, which represents the point of minimum solubility of the crystal nucleus (i.e., solute) in the oil (i.e., solvent). The temperature at which this point is achieved depends on the extent of intermolecular interactions between the TG of highest melting T° and the rest of the TG in the oil system, these interactions occurring at the level of the bilayer lamellar structures. Thus, the higher the interactions (i.e., similar or complementary conformations of different families of TG's), the lower the T° needed to accomplish the point where $\eta_{red} = 0$, making the viscosity of the liquid phase a limiting factor for the growth of the crystal nucleus, since diffusion of the crystallizing molecule is an inverse function of viscosity (33).

Nucleation mainly depends on the extent of supercooling/supersaturation, the thermodynamic driving force for crystallization. In contrast, crystal growth is affected by several factors being the more important diffusion of molecules to the surface of growing crystals and incorporation of molecules on the crystal surface (33). Avrami's crystalliza-tion rate constant, z, is a composite constant incorporating nucleation and crystal growth rate characteristics (59-61) (Table II), and this rate constant has been utilized to study the particular effect of oil viscosity and supercooling on TG crystallization in a system form by tripalmitin in sesame oil. This particular system was chosen since represents a model vegetable oil in which a high melting temperature TG (i.e., tripalmitin) is in solution with unsaturated TG's (i.e., sesame oil). In this study z was calculated with a modified Avrami equation as proposed by Khanna and Taylor (60), since it has been shown that evaluation of crystallization rate constant and associated energy of activation, Ea, using Avrami's original equation leads to erroneous results. These erroneous results are consequence of the influence of n, the Avrami index, on the magnitude of z (60). Equations 10 and 11 show the original and modified Avrami model for spherulite crystallization, where F is the fractional TG crystallization as a function of time, t. The magnitude of F may be calculated as $F = (T_i - T)/(T_i - T_f)$ where T_i is the concentration of solid (i.e., crystal) at time zero, T is the concentration of solid at time t, and T_f is the maximum concentration of solid obtained in the crystallization process. These values might be determinated by transmittance (26) or DSC measurements (61).

$$-\text{Ln}(1 - F) = zt^n \tag{10}$$

$$-\text{Ln}(1 - F) = (zt)^n \tag{11}$$

Table II. The n and z constants in Avrami's equation (adapted from ref. 61).

Mechanism of crystal growth	Nucleation sporadic in time (Primary nucleation)	
	n	z
Polyhedral	4	$\pi G^3(I/3)$
Plate-like	3	$\pi G^2(I/3)$
Cylindrical	2	$\pi d^2 G(I/r)$

n, index of Avrami's equation; z, rate constant in Avrami's equation; G, linear growth rate of crystal sperulite; I, sporadic nucleation rate in time; d, width of crystal fibril; r, crystal radii.

From the above data it is evident that the Ea associated to z represents the energy involved in the overall reaction for crystallization as affected by supercooling/supersaturation, molecular diffusion (i.e., viscosity), or incorporation of the crystallizing molecule on the crystal surface. In the case of TG crystallization from the melt in vegetable oils, one of these factors predominates and limits or favors the rate of TG nucleation or its crystal growth rate. To investigate these phenomena in vegetable oil crystallization we have evaluated crystallization utilizing the following conditions: (a) constant viscosity and different degrees of supercooling, and (b) constant supercooling and modifying the magnitude of viscosity. In these experiments the same oil model system of tripalmitin solutions in sesame oil at different concentrations was utilized. Table III shows the respective z and Ea obtained in such crystallization conditions. Figure 9 shows the magnitude of the oil viscosity, parameter inversely associated with molecular diffusion, as a function of supercooling for the different tripalmitin solutions in sesame oil (62). This graph indicates the viscosity of the liquid phase in which tripalmitin molecules had to diffuse during nucleations and further crystal growth. The results of these experiments indicated that, independent of tripalmitin concentration, the size of tripalmitin crystals followed an inverse relationship with the viscosity of the liquid phase in which crystallization took place. This, provided similar supercooling conditions (e.g., 22.0 - 22.5°C, Figure 9) and equivalent time for crystallization (i.e., 30 min after induction time for nucleation) were utilized. In contrast, nucleation followed direct relationships with the degree of supercooling provided a constant viscosity (e.g., 5.25 - 5.5 dynes/cm^2) and equal crystallization time were utilized.

In conclusion, during TG crystallization from the melt the concentration of the TG with the highest melting temperature in the vegetable oil determines the degree of supercooling required to achieve crystallization. In consequence, for different lots of vegetable oil (i.e., with different concentration of the TG of highest melting temperature) the magnitudes of z and the crystallization Ea are more associated with the crystal growth process as viscosity of the oil decreased. In comparison, as viscosity of the oil increases the main crystallization process evaluated through the magnitude of z and Ea is nucleation. On the other hand, when viscosity is constant among different lots of the oil the increase in supercooling results in a higher degree of nucleation without an appreciable effect on crystal size. The crystallization conditions above described determine which event,

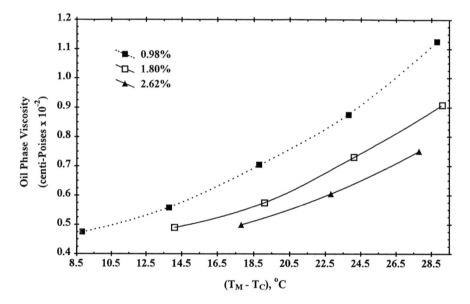

Figure 9. Viscosity of tripalmitin sesame oil solutions (shear rate utilized, 7.92 s⁻¹) as a function of supercooling. T_M was 311.8 K for the 0.98% solution, 317.1 K for the 1.80 % solution, and 320.8 K for the 2.62% solution (adapted from ref. 62).

nucleation or crystal growth, predominates. In turn this determines the final rheological profile of the crystal suspension. The rheology of a crystal network in which a liquid phase is trapped (i.e., fractal structure), and its time/temperature dependencies (e.g., TG polymorph transformations) are important factors that determine processing variables and final product characteristics of food like butter, margarine, chocolate, and spreads.

Table III. Crystallization Rate Constant, z, Calculated with the Modified Avrami Equation (equation 11) and Associated Energy of Activation (Ea) for the Crystallization Process of Tripalmitin in Sesame Oil at Three Concentrations (adapted from ref. 62).

Temperature (K)	Tripalmitin concentration (%, w/v)	z (min^{-1})	Ea (kCal/mol)
287.0	0.98	0.04420	45.97
289.0		0.03367	
291.0		0.02578	
293.0		0.01104	
294.0		0.00706	
295.0		0.00552	
294.0	1.80	0.02638	34.75
295.0		0.02248	
296.0		0.01935	
297.0		0.01349	
298.0		0.01122	
299.0		0.01069	
294.5	2.62	0.07065	48.81
295.5		0.04954	
296.0		0.02993	
297.5		0.02678	
298.5		0.01910	
299.0		0.01964	

The Future in Vegetable Oils Demand and Research

The functional properties are not the only aspects that determine oil/fat acceptance for today's consumer. The health effect of dietary lipids is getting constant attention from the consumer and now is a major concern affecting oil and fat consumption patterns. Consumers are now aware that a diet high in saturated fatty acids is extensively associated to a cholesterol raising effect increasing the risk of arteriosclerosis-related problems. Multiple studies have demonstrated this effect beyond doubt in studies carried out in carefully controlled metabolic investigations (63-65). In humans this cholesterol raising effect of saturated fatty acids occurs independently of dietary cholesterol. However, the exact mechanism how this occurs is still under investigation (66). These, and other health related claims associated with the dietary intake of lipids, have produced that consumers shift their consumption patter from animal fats toward vegetable oils, and in some segments of consumer population, the look for fat elimination from the diet. This last point is not advisable since from a nutritional point of view fats and oils represent essential dietary components for human health.

We are still learning how the complex enzymatic systems determine triacylglyceride hydrolyzation in the human digestive tract. Today, we know that not all fatty acids have the same destiny, function, and significance from the nutritional and physiological point of view (66). The absorption, transport and final destination of fatty acids depend on their physical properties (i.e., melting temperature, solubility), associated chemical characteristics (i.e., degree of unsaturation, length of the fatty acid chain), and stereospecific distribution in the glycerol. This knowledge is now applied in the design of oils utilized for pharmaceutical, oleochemical, and food purposes.

In the particular case of designed oils with applications in food and medical products, researchers and companies are intensively exploring the use of structured lipids. Structured lipids are TG's containing mixtures of short, medium, and long chain fatty acids attached to the glycerol backbone for specific functionality (67). Structured lipids with particular compositions might only be produced through enzymatic interesterification of oils or genetic manipulation of the vegetables. Other techniques, like traditional chemical interesterification, produce randomized structured lipids with variable physical chemical properties and therefore inconsistent functionality.

Genetic engineering, weather standard cross breeding or transgenic techniques, has been used successfully to modify vegetable oil properties and composition. For instance, sunflower oil, with its naturally low linolenic acid concentration, is now marketed in a new high oleic version competing for the consumer segment appealed for the "healthy effect" of the Mediterranean diet, which lipid source is mainly based on olive oil (55-83% oleic acid) (68, 69). In the same way, Calgene is marketing its high laurate canola oil under the name Laurical. This oil is composed of structured triacylglycerides created by genetically inserting the lauryl-APC thioesterase enzyme from the seed of the California bay tree into canola, which does not contain lauric fatty acids. This oil, given its melting/crystallization profile, is now used in confectionery, coatings, coffee whiteners, whipped toppings, and filling fats (67). These and other commercial products like Salatrim, a family of short and long structured triacylglyceride molecules which provide about 5 calories/g, are examples of new oils that are expect to fill the usually contradictory consumer needs for nutritional, healthy, and functional oil products (67, 70, 71). The future for vegetable oils production seems to be directed to continue the efforts of plant breeders in providing oils with modified composition, targeting enhanced functional properties with healthier fatty acids profiles. However, research costs and regulatory issues are the main obstacles that limit the widespread use of structured lipids derived from genetically modified vegetable sources in food products.

Literature Cited

1. Christie, W.W. *High performance liquid chromatography and lipids*; Pergamon Books, Oxford, **1987**.
2. MacRitchie, F. *J. Sci. Food Agric.* **1977**, *28*, 53-58.
3. Takahashi, S.; Seib, P.A. *Cereal Chem.* **1988**, *65*, 474-483.
4. Morrison, W.R. *J. Cereal Sci.* **1988**, *8*, 1- 15.
5. Godet, M.C.; Tran, V.; Colonna, P.; Buleon, A.; Pezot, M. *Int. J. Biol. Macromol.*, **1995**, *17*, 405-408.
6. Chobert, J.M.; Bertrand-Harb, C.; Nicolas, M.G. *J. Agric. Food Chem.* **1988**, *36*, 883-892.
7. Yamauchi, K.; Shiizu, M.; Kamiya, T. *J. Food Sci.* **1980**, *45*, 1237- 1242.
8. Graham, D.E.; Phillips, M.C. *J. Colloid Inter. Sci.* **1980**, *76*, 240-250.
9. Toro-Vazquez, J.F.; Regenstein, J.M. *J. Food Sci.* **1989**, *54*, 1177-1185.
10. Izzo, M.T.; Ho, Ch. *Cereal Chem.* **1989**, *66*, 47-51.
11. Larsson, K. *Lipids - molecular organization, physical functions and technical applications*; The Oily Press Ltd, West Ferry, Dundee, **1994**, Vol, 5.
12. Larsson, K. *Trends Food Sci Technol.* **1994**, *5*, 311-315.
13. Davis,T.R.; Dimick, P.S. *J. Am. Oil Chem. Soc.* **1989**, *66*, 1494-1498.
14. Gordon, M.H.; Abdul, R. I. *J. Am. Oil Chem. Soc.* **1991**, *68*, 577-579.
15. deMan, J.; deMan, L. *Malaysian Oil Sci. Technol.* **1995**, *4*, 56-60.
16. Fitch, H.B. *INFORM.* **1992**, *3*, 1172-11178.
17. Boyle, E. *Food Technol.* **1997**, *51(8)*, 52-59.
18. Baldwin, E.A.; Nisperos, M.O.; Hagenmaier, R.D., Baker, R.A. *Food Technol.* **1997**, *51(6)*, 56-64.
19. Nawar, W.W. In *Food Chemistry*; Fenema, O.R., Ed.; Marcel Dekker, New York, **1996**, pp *225-319*.
20. Blumenthal, M.M. *Food Technol.* **1991**, *45*(2), 68-71.
21. Dutt, N.V.K.; Prassad, D.H.L. *J. Am. Oil Chem. Soc.* **1989**, *66*, 701-703.
22. Noureddini, H.; Teoh, B.C.; Clements, L.D. *J. Am. Oil Chem. Soc.* **1992**, *69*, 1189-1191.
23. Toro-Vazquez, J.F.; Mendez-Montealvo, G. *J. Am. Oil Chem. Soc.* **1995**, *72*, 675-679.
24. Toro-Vazquez, J.F.; Rocha-Uribe, A. *J. Am. Oil Chem. Soc.* **1993**, *70*, 589-594.
25. Toro-Vazquez, J.F.; Gallegos-Infante, A. *J. Am. Oil Chem. Soc.* **1996**, *73*, 1237-1246.
26. Dibildox-Alvarado, E.; Toro-Vazquez, J. *J. Am. Oil Chem. Soc.* **1997**, *74*, 69-76.
27. Lius, H.; Przybylski, R.; Eskin, N.A.M. *J. Am. Oil Chem. Soc.* **1996**, *73*, 1137-1141.
28. Dutt, N.V.K.; Ravi-Kumar, Y.V.L.; Samanti-Vedanayagam, H. *J. Am. Oil Chem. Soc.* **1991**, *69*, 1263.
29. Toro-Vazquez, J.F.; Infante-Guerrero, R. *J. Am. Oil Chem. Soc.* **1993**, *70*, 1115-1119.
30. Noureddini, H.; Teoh, B.C.; Clements, L.D. *J. Am. Oil Chem. Soc.* **1992**, *69*, 1189-1191.

252

31. Larsson, K. *Fette. Seifen. Anstrich.* **1972**, *74*, 136-142.
32. Bohdanecký, M.; Kovàr, J. *Visosity of polymer solutions*; Elsevier Scientific Publishing Co., New York, **1982**.
33. Hartel, R.W. In *Physical chemistry of foods*; Schwartzber, H.G. and Hartel, R.W., Eds.; Marcel Dekker, New York, **1992**, pp 47-81.
34. Cavanagh, G.C. In *World conference proceedings in edible fats and oil processing, basic principles and modern practices*; Erickson, D.R., Ed.; American Oil Chemists' Society, Champaign, **1990**, pp 107-116.
35. Vreeker, R.; Hoekstra, L.L.; den Boer, D.C.; Agterof, W.G.M. In *Food colloids and polymers: stability and mechanical properties*; Dickinson, E. and Walstra, P., Eds.; The Royal Society of Chemistry, Cambrige, **1993**, pp 15-22.
36. Breitschuh, B.; Windhab, E.J. *J. Am. Oil Chem. Soc.* **1996**, *73*, 1603-1610.
37. Krautwurst, J. *Kieler Milchwirtschaftliche Forschungsberichte*, **1970**, *22*, 255-284.
38. Sherbon, J.W.; Dolby, R.M. *J. Dairy. Sci.* **1973**, *56*, 52-60.
39. *Official Methods and Recomended Practices of the American Oil Chemists'*, The American Oil Chemists' Society, Champaign, Illinois, **1994**.
40. Norton, I.T.; Lee-Tuffnell, C.D.; Ablett, S.; Bociek, S.M. *J. Am. Oil Chem. Soc.*, **1985**, *62*, 1237-1244.
41. Malsen, K.V.; Peschar, R.; Schenk, H. *J. Am. Oil Chem. Soc.*, **1996**, *73*, 1209-1215.
42. Blaurock, K. *INFORM*, **1993**, *4*, 254-259.
43. Minato, A.; Ueno, S.; Yano, J.; Wang, Z.H.; Seto, H.; Amemiya, Y.; Sato, K. *J. Am. Oil Chem. Soc.*, **1996**, *73*, 1567-1572.
44. Kellens, M.; Meeussen, W.; Riekel, C.; Reynaers, H. *Chem, Phys. Lipids*, **1990**, *52*, 79-98.
45. Kellens, M.; Meeussen, W.; Hammersley, A.; Reynaers, H. *Chem, Phys. Lipids*, **1991**, *58*, 145-158.
46. Meakin, P. *Adv. Colloid Interface Sci.* **1988**, 28, 249-275.
47. Mensink, R.P.; Katan, M.B. *N. Engl. J. Med.* **1990**, *323*, 439-445.
48. Hasman, J.M. *INFORM*, **1995**, *6*, 1206-1213.
49. Fitch Haumann, B. *INFORM*, **1994**, *5*, 668-678.
50. Banks, W.; Clapperton, J.L.; Girdler, A.K. *J. Sci. Food Agric.* **1985**, *36*, 421-432.
51. Kaylegian, K.E.; Lindsay, R.C. *J. Dairy Sci.* **1992**, *75*, 3307-3317.
52. Hamm, W. *Tends Food Sci. Technol.* **1995**, *6, 121-126.*
53. Liu, H.; Przybylski, R.; Eskin, N.A.M. *J. Am. Oil Chem.* **1996**, *73*, 1557-1566.
54. Pantzaris, T.P. *Pocketbook of palm oil uses*; Palm Oil Research Institute of Malaysia, Kuala Lumpur, **1994**.
55. Basiron, Y.; Abdullah, R. *INFORM,* **1994**, *6*, 891-894.
56. *Palm Oil Technical Bulletin*, **1997**, Vol. 3, No 1,.Palm Oil Research Institute of Malaysia, Kuala Lumpur.
57. Boistelle, R. In *Fundamentals of nucleation and crystal growth*; Garti, N., Sato, K., Eds.; Surfactant Science Series; Marcel Dekker, Inc., New York, **1988**, Vol. 31; pp 189-226.
58. Mersmann, A. *Crystallization technology handbook*; Marcel Dekker, New York, **1995**.

59. Avrami, M. *J. Chem. Phy.* **1940**, *8*, 212-224.
60. Khanna, Y.P.; Taylor, J. *J. Pol. Eng. Sci.* **1988**, *28*, 1042-1045.
61. Kawamura, K. *J. Am. Oil Chem. Soc.* **1979**, *56*, 753-758.
62. Toro-Vazquez, J.F.; Dibildox-Alvarado, E. *J. Food Lipids*, **1997**, *4*, 269-282.
63. Hegsted D.M. In *Health Effects of Dietary Fatty Acids*; Nelson, G.J., Ed.; American Oil Chemists´ Society, Champaign, IL.,**1990**, pp 50-68.
64. Mensink N, R.; Katan, M.B. *N. Engl. J. Med.* **1990**, *323*, 439-445.
65. Zock, P.L.; Katan, M.B. *J. Lipids Res.* **1992**, *33*, 399-410.
66. Valenzuela, A.; Sanhueza, J.; Nieta, S. *A and G*, **1997**, *29*, 582-588.
67. Fitch, H.B. *INFORM*, **1997**, *8*, 1004-1011.
68. Fitch, H.B. *INFORM*, **1996**,*7*, 890-903.
69. Reddy, B.S. In *Health Effects of Dietary Fatty Acids;* Nelson, G.J., Ed.; American Oil Chemists´ Society, Champaign, IL., **1990**, pp 157-166.
70. Giese, J. *Food Technol.*, **1996**, *50* (4), 78-83.
71. Deis, R.C. *Food Product Design*, **1997**, Nov., 29-55.

Chapter 16

Low Calorie Fats and Sugar Esters

Casimir C. Akoh

Department of Food Science and Technology, The University of Georgia, Athens, GA 30602–7610

Low calorie fats and sugar esters are synthetic fats often known as fat substitutes intended to replace conventional fats and oils in food formulations and preparations. They are made by esterification of the glycerol or sugar moiety with fatty acids. Depending on the degree of esterification, they may serve as a low, reduced or zero calorie fat substitutes. Whether they contribute calorie or not depends on their susceptibility to hydrolysis by the lipolytic enzymes. These fats can replace most of the physical, functional and organoleptic properties of conventional fats and oils in foods. Olestra, a sucrose fatty acid polyester, is the only synthetic zero calorie fat substitute approved for use in savory snacks in the United States of America to date. Several other molecules with similar properties are under development or developed but not yet approved for food use.

Fats and oils contribute to the palatability and satiety of foods. High intake of dietary fat is associated with increased risk of obesity, some types of cancer and sometimes gallbladder disease. There is a strong epidemiological evidence linking high consumption of saturated fat to high serum cholesterol and increased risk for coronary heart disease (*1,2*). Current recommendation is to limit total fat intake to no more than 30% of daily energy intake, with monounsaturated and polyunsaturated fats accounting for at least two-thirds of this intake and saturated fats no more than 10% (*3*). Consumers are showing interest in reducing dietary calories from fat. At the same time, they feel reluctant to give up their dietary habits if taste is compromised because their acceptance of any food product depends largely on the taste. The food industry in response to consumer demand is developing alternatives to full fat foods in the form of reduced, low, or zero calorie fats with the potential to partially or fully replace the calories from fat.

Low Calorie Fats Based on Sugar Moiety

Olestra. The discovery of nondigestible sucrose fatty acid polyesters, better known as olestra, while studying the digestion and absorption of fats led to the development of low calorie fats (*4*). Olestra was later developed by Procter and Gamble Co. (Cincinnati, OH) who petitioned the United States Food and Drug Administration (FDA) and obtained approval in January 1996 (*5*) for its use in replacing up to 100% of the conventional fat in savory snacks such as potato chips, cheese puffs, and crackers; for frying of savory snacks; and for use as dough conditioners and oil sprays for flavor.

The process for the manufacture of olestra has been described in detail elsewhere (6-8). Briefly, both saturated and unsaturated fatty acids methyl esters (FAME) of chain length C12-C24 are obtained from vegetable oils and animal fats by chemical hydrolysis and methylation. The FAME are added to sucrose or sucrose octaacetate and the mixture transesterified by alkali metals or their soaps, under anhydrous conditions for 2-16 h depending on the reactants.The crude product is purified by washing with warm water and alcohol, bleached and deodorized to remove free fatty acids and odor materials. Distillation is performed to remove unreacted FAME and sucrose fatty acid esters with low degrees of esterification or substitution with fatty acids. The purified product is known as sucrose fatty acid polyester esterified with 6-8 fatty acids. The structure of olestra is shown in Figure 1.

Olestra resembles conventional fats and oils in physical properties and appearance. The properties of a particular product depends on the type of fatty acids used in the synthesis process and the degree of substitution of sucrose hydroxyl or acetate groups with long chain fatty acids. Olestra is stable to cooking and frying temperatures and may be used in food formulations such as in baked goods, salad dressing, frozen desserts, margarine, shortening, spread and butter, dairy products, processed meats, snack products, confectioneries, soups, sauces and gravies for various functionalities such as heat conduction, emulsification, texturizing, increased viscosity and spreadability.

Olestra is neither digested nor absorbed in the gastrointestinal track. However, it can affect the absorption of certain fat soluble vitamins and nutrients that may partition into it. It has the potential to cause some gastrointestinal discomfort such as abdominal cramps and loose stools. The FDA requires that foods containing olestra be labeled with the statement "This Product contains Olestra. Olestra may cause abdominal cramping and loose stools. Olestra inhibits the absorption of some vitamins and other nutrients. Vitamins A, D, E, and K have been added" (5). Supplementation with vitamin A compensates for olestra's effect on the provitamin A function of carotenoids. Olestra does not significantly affect the absorption of other macronutrients such as carbohydrates, triacylglycerols, proteins, or water soluble vitamins and minerals. The Center for Science in the Public Interest (CSPI, Washington, DC), an advocacy and education organization, opposed the approval of olestra and petitioned the FDA to repeal the approval on the basis of several alleged adverse effects, including gastrointestinal discomfort. Olestra may lower serum lipids in some individuals, help the obese lose weight, and benefit persons at high risk of coronary heart disease and colon cancer patients (9-11). Sources of selected low calorie fats are shown in Table I.

Polyol Fatty Acid Polyesters.   Other polyol fatty acid polyesters with potential for use in place of sucrose polyesters as low calorie fats and oils are the monosaccharides (mannitol, sorbitol), disaccharides (lactitol, trehalose), trisaccharide (raffinose), and tetrasaccharide (stachyose) fatty acid polyesters. If the degree of the polyol hydroxyl group substitution with fatty acids is greater than 4, the susceptibility of the polyol fatty acid polyester to lipolysis decreases. Their synthesis, purification, analysis and physical properties are essentially the same as that of sucrose fatty acid polyesters (6). None of these molecules is approved for use in foods by the FDA. However, further studies and product developments will be required before petitions can be filed. With olestra paving the way, it is expected that their approval as olestra competitors will take less time.

Sorbestrin.   Sorbestrin, a mixture of tri-, tetra-, and pentaesters of sorbitol and sorbitol anhydride with fatty acids was developed by Pfizer, Inc. (now owned by Cultor Food Science, New York, NY). Its caloric value is 1.5 kcal/g. Sorbestrin is heat stable

Table I. Sources of Some Low Calorie Fats

Name(s)	Composition/description	Potential Source
Olestra/olean	sucrose fatty acid polyester	Procter & Gamble Co. (Cincinnati, OH)
Sorbestrin	sorbitol and sorbitol anhydride fatty acid polyester	Cultor Food Science (New York, NY)
Alkyl glycoside polyesters	alkyl glycoside fatty acid polyesters	Curtice Burns, Inc. (Rochester, NY), under development
MCT	medium chain triacylglycerols	Stepan Co. (Maywood, NJ) and Abitec Corp. (Columbus, OH)
Laurical	high laurate canola oil (SL)	Calgene, Inc. (Davis, CA)
Betapol	structured lipid that mimics human milk with C16:0 at sn-2 position of glycerol, enzyme produced (SL)	Loders Croklaan (Netherlands) and Unilever (UK). Available in Europe
Caprenin	caprocaprylobehenin (SL)	Procter & Gamble Co. (Cincinnati, OH)
Benefat/salatrim	mixture of C2:0-C4:0 & C18:0 (SL)	Cultor Food Science (New, NY) and Nabisco Foods Group (East Hanover, NJ)
Neobee SL	structured lipids (SL) of varying fatty acid compositions	Stepan Co. (Maywood, NJ)
Captex	structured lipids of varying fatty acid compositions	Abitec Corp. (Columbus, OH)
EPGs	esterified propoxylated glycerols	ARCO Chemical Co. (Wilmington, DE) and CPC International/Best Foods (Englewood Cliffs, NJ)
PGEs	fatty acid esters of polyglycerol	Lonza, Inc. (Fair Lawn, NJ)

and can be used as low calorie fat for frying, salad dressing and baked goods. It is not commercially available yet.

Alkyl Glycosides Fatty Acid Polyesters. Alkyl glycosides suitable for the synthesis of fatty acid polyesters include methyl or ethyl glucose and galactose or octyl-ß-glucose. Among these, methyl glucoside fatty acid polyester has received greater attention as a possible replacement of conventional fats and oils in foods. The preparation of alkyl glycoside fatty acid polyesters and their physical properties have been described elsewhere (*6, 12*). In general, up to 4-5 hydroxyl groups of the glycosides are esterified with fatty acids. Alkyl glycoside fatty acid polyesters can be used as frying oils and in making Italian or white salad dressings. Curtice Burns, Inc. (Rochester, NY) has the patent on the application of alkyl glycoside fatty acid polyesters (*13*). No approval petition has been filed with the FDA.

Low or Reduced Calorie Fats Based on Glycerol Moiety

The fact that some saturated fatty acids are poorly absorbed, especially when they are at specific positions on the glycerol molecule coupled with the differences in caloric values of short, medium, and long chain fatty acids, led to the development of a new group of lipids or fat substitutes that can be used as low or reduced calorie fats and oils. For example the caloric availability (kcal/g) of acetic acid = 3.5, propionic = 5.0, butyric = 6.0, long chain fatty acid (C14:0-C22:0) = 9.5, while glycerol = 4.3. The triacylglycerols can be hydrolyzed by the lipolytic enzymes and do supply some calories. Among these are the medium chain triacylglycerols (MCTs) and structured lipids (SL).

Medium Chain Triacylglycerols. These contain predominantly saturated fatty acids (C8:0-C10:0) with traces of C6:0 and C12:0. The fatty acids are obtained from coconut, palm kernel, cohune, and tocum vegetable oils after hydrolysis and concentration. MCTs are then prepared by reesterification of the fatty acids with glycerol to form triacylglycerols (*14,15*).Their caloric value is approximately 8.3 kcal/g. MCTs are oxidatively stable and are more soluble in water than conventional long chain triacylglycerols. They are commercially available on the basis of GRAS self determination and can be used as a solvent or carrier for flavors, carrier for colors and vitamins, and as confectionery coatings to provide gloss and prevent sticking. Medically, they have been used since the 1950s in enteral and parenteral diets for persons with lipid absorption, digestion, or transport disorders, because they are metabolized differently than long chain triacylglycerols (LCT). MCT can be used by AIDS, cystic fibrosis, multiple trauma, burn injury, and critically ill patients who may be hypermetabolic due to trauma and or/sepsis (*22*). They are a source of easily absorbed fat and rapidly utilized energy (*16*).

Structured Lipids. These are triacylglycerols containing short chain fatty acids (SCFA) and/or medium chain fatty acids (MCFA), and long chain fatty acids (LCFA) esterified to glycerol molecules. They are manufactured by chemical or enzymatic transesterification (*17-19*) for specific purposes, such as a reduced calorie fat in foods or for medical applications (*20*) or for infant formula for premature infants. SL have modified absorption rates and may be useful in improving fat utilization in patients with cystic fibrosis or pancreatic and fat malabsorption disorder. Figure 2 shows the general structure of SL. Recently, Calgene Inc. (Davis, CA) used antisense technology to produce non-synthetic canola oil-based structured lipids (SL) with the MCFA at the *sn*-1 and *sn*-3 positions, and LCFA at the *sn*-2 position (*21, 22*). This product is marketed under the trade name Laurical (high laurate canola). All synthetic SL involve random placement of fatty acids on the glycerol backbone. However, enzymatic synthesis

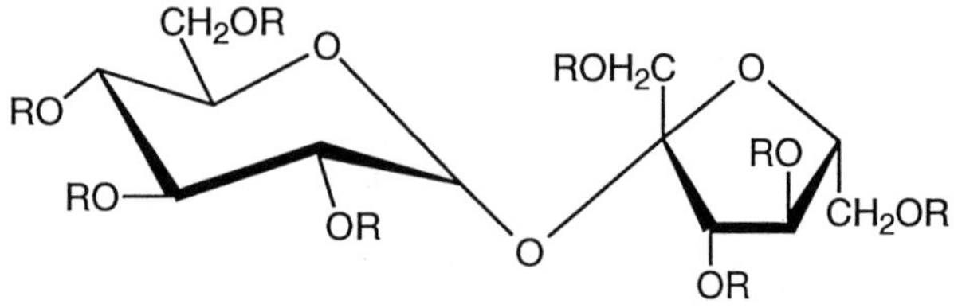

Sucrose Polyester (Olestra®, or Olean®) [α Glucopyranosyl-
$(1 \rightarrow 2)$ - β Fructofuranoside linkage]

Where R = Acyl group of fatty acids $\left(R\text{-}\overset{\overset{O}{\|}}{C}\text{-} \right)$

Figure 1. Structure of sucrose fatty acid polyester, olestra.

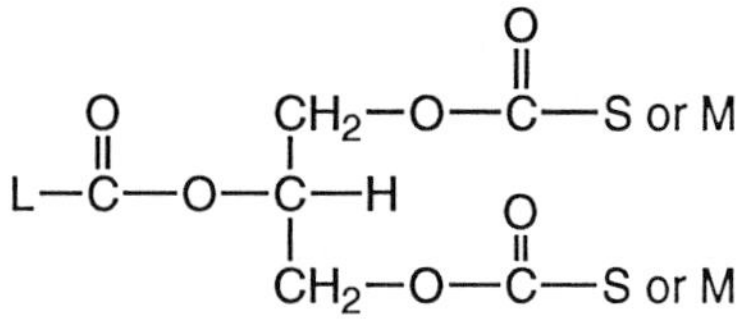

General Structure of Structured Lipids

Where: S or M or L is from short-chain,
medium-chain, or long-chain fatty acid,
respectively. The position of S or M or
L is interchangeable.

Figure 2. General structure of structured lipids.

allows the production of more specialized SL in which specific fatty acids are placed at specific positions on the glycerol backbone (*19*). The only commercially available SL produced enzymatically is "Betapol" (Unilever, UK) used in infant formula in Europe to mimic the human milk fatty acids. Approximately 60% of the palmitic acid in Betapol is at the *sn*-2 position for improved absorption. Loders Croklaan (Netherlands) a unit of Unilever is currently producing and marketing Betapol. Several molecular species of SL targeted for nutrition, foods, consumer products and medical applications are being synthesized enzymatically at various universities and industries worldwide. Price will be a key determining factor in the commercialization of the enzymatic process and the use of the products in foods. Some of the commercially available synthetic SL are described below.

Caprenin. Caprocaprylobehenin, commonly known as caprenin, is a reduced calorie SL derived from the random esterification of glycerol with caprylic (C8:0), capric (C10:0), and behenic acid (C22:0). Caprenin was developed by Procter & Gamble Co. Because behenic acid is partially absorbed and caprylic and capric acids are more easily metabolized than other LCFA, caprenin is only 5 kcal/g. Its functional properties resemble cocoa butter and can be used in soft candy and confectionery coatings. A GRAS affirmation petition has been filed by Procter & Gamble Co. for use in soft candy and confectionery coatings (*23*).

Benefat. Benefat, a brand name for salatrim (short and long-chain acid triglyceride molecule) is a family of structured triacylglycerols originally developed by Nabisco Foods Group and now marketed by Cultor Food Science (New York, NY) as a reduced calorie SL with a caloric value of 5 kcal/g. It is composed of a mixture of SCFA (C2:0-C4:0, acetic to butyric acids) and LCFA (predominantly C18:0, stearic acid). The SCFA are chemically transesterified with vegetable oils such as highly hydrogenated canola or soybean oil. Stearic acid is partially absorbed while the SCFA contribute only a few calories (*24*). It has the taste, texture and functional properties of conventional fats and oils. Benefat I was developed primarily for cocoa butter replacement. Benefat is not suitable for deep fat frying because the short chain fatty acid components may volatilize. Benefat can be produced to have different melting profiles by adjusting the amounts of SFCA and LCFA used in their chemical synthesis. Products in the market that contain Benefat are the reduced fat chocolate-flavored baking chips introduced in 1995 by the Hershey Food Corporation and SnackWell's fudge-dipped granola bars. Benefat can also be used as a cocoa butter substitute. Benefat is intended for use in chocolate-flavored coatings, deposited chips, caramel, fillings for confectionery and baked goods, margarines and spreads, savory dressings, dips and sauces and in dairy products (*25*). FDA accepted Nabisco's GRAS petition for filing in 1994.

Neobee and Captex Series SL. The Neobee series of structured lipids are manufactured chemically by interesterification and marketed by Stepan Co. (Maywood, NJ). These SL are tailor-made for customers and applications such as in medicine. For example, Neobee-1814 is made by random interesterification of MCT and butter oil (50:50, w/w). Other products from Stepan are SL-110 (interesterified MCT and soybean oil), SL-120 (MCT and menhaden oil), SL-130 (MCT and sunflower oil), SL-140 (MCT, menhaden, soybean oils and tributyrin), SL-210 (coconut and soybean oil), SL-220 (coconut, menhaden and canola oils), SL-230 (coconut, menhaden and soybean oils and tributyrin), SL-310 (MCT and menhaden oil), and SL-410 (MCT, butter and sunflower oils). Abitec Corp. (Columbus, OH) makes and markets the synthetic Captex series of SL. Examples include Captex-350 (transesterified coconut oil with caprylic and capric acids) and 810D (contains caprylic to linoleic acids). The main MCFA in the Neobee and Captex products are caprylic, capric and lauric acids while the LCFA include oleic, linoleic, eicosapentaenoic and docosahexaenoic acid. Uses include nutritional supplements, clinical and consumer products. Their use in food applications have not received adequate attention.

Esterified Propoxylated Glycerols (EPGs). These are derivatives of propylene oxide, synthesized by reacting glycerol with propylene oxide to a polyether polyol. This polyether polyol is subsequently esterified with desired fatty acids. EPG is different from conventional triacylglycerols because they contain an oxypropylene group between the glycerol backbone and the fatty acids. ARCO Chemical Co. (Wilmington, DE) has the patent on EPG. Currently, EPG is being developed by ARCO Chemical Co. and CPC International/Best Foods (Englewood Cliffs, NJ) as a replacement for fats and oils in frozen desserts, salad dressings, baked goods, and spreads and for cooking and frying. Because they are resistant to lipolysis by lipases, they have low caloric value.

Polyglycerol Esters (PGEs). PGEs are hybrid fats with fatty acid side chains and glycerol backbone synthesized with alkali catalysts at 230°C. They are used as emulsifiers, fat substitutes, surfactants in whipped toppings and frozen desserts, stabilizers in beverages, and texture enhancers in cake mixes. The caloric value is approximately 6-7 kcal/g (26).

Sugar Esters

Sugar ester is a term used to describe all fatty acid esters of mono- through polysaccharides including polyols and alkyl glycosides, but with a low degree of esterification or substitution with fatty acids (typically, 1-4 fatty acids substitutions). The most studied is sucrose fatty acid esters (SFEs).

Sucrose Fatty Acid Esters. SFEs are nonionic surface active agents. They are tasteless, odorless and nontoxic. They are easily hydrolyzed by lipases and are absorbable. SFEs are biodegradable and are suitable for food, pharmaceutical and cosmetic applications. SFEs are approved for use in foods in Japan, East Asia, United States and Europe.

Properties and production of SFEs were reviewed recently (6, 27). SFE can be manufactured by direct esterification of sucrose and fatty acyl chloride or anhydride, transesterification of fatty acid ester and sucrose, interesterification between acetylated sucrose and fatty acid ester, or by lipase catalyzed reactions. The preferred industrial method is the solvent-free transesterification between sucrose and fatty acid ester with appropriate chemical catalyst. The process leads to randomization with no preference of the fatty acids to any hydroxyl position. However, the enzymatic approach, though not commercialized can lead to selective acylation of the primary hydroxyl group and high monoester products. SFEs can be isolated by purification which recently involves washing with water and avoidance of organic solvents for food grade products (27).

The structure of sucrose monoester is shown in Figure 3. They have a wide range of hydrophilic-lipophilic balance (HLB value 1-19) depending on the degree of esterification of the free hydroxyl groups in sucrose (Table II). SFEs are suitable for applications in o/w and w/o food systems. They function by reducing the surface tension between water and oil surfaces. The choice of a particular SFE will depend on the application and is easily achieved by proper experimentation. Their main function in foods is emulsification, antimicrobial agent, fat crystal inhibition and retrogradation of starch. SFEs are capable of complexing with starch, proteins, and monoacylglycerols. In baked goods, they increase loaf and cake volumes and act as shortening distribution agent as well as prevent retrogradation and staling of bread. They can be used to improve cookie spread factor. Generally, the SFEs with higher HLB values are more effective than lower HLB esters in improving loaf volume, cookie spread factor and

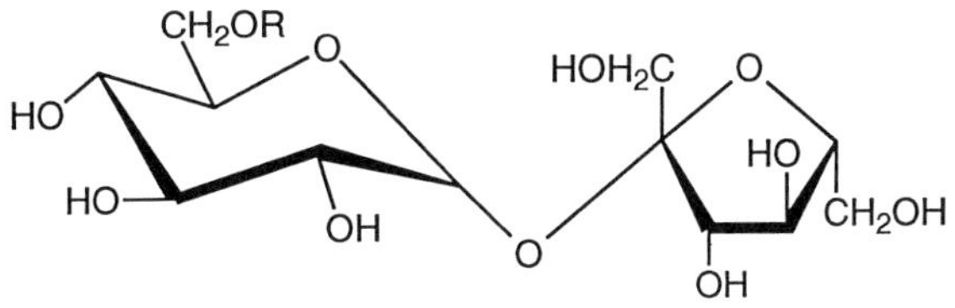

Sucrose Monoester (sucrose fatty acid ester)

Where R = Acyl group of fatty acids $\left(R-\overset{O}{\underset{||}{C}}- \right)$

Figure 3. Structure of sucrose monoester.

Table II. Commercial Sources and Properties of Sugar Esters

Ryoto Esters[a]	DK Esters[b]	Approximate HLB[c]	Ester Content (%) Monoester	Form Di-/Tri-	Polyester
S-170	F-10	1	0	100	powder
S-270	F-20W	2	10	90	powder
S-370	-	3	20	80	powder
S-570	-	5	30	70	powder
	F-50	6	30	70	powder
S-770	-	7	40	60	powder
	F-70	8	40	60	powder
S-970	-	9	50	50	powder
	F-90	9.5	45	55	powder
S-1170	F-110	11	55, 50	45, 50	powder
	F-140	13	60	40	powder
S-1570	F-160	15	70	30	powder
S-1670	-	16	75	25	powder
-	SS	19	100	0	powder

[a]Ryoto Sugar Esters, produced by Mitsubishi Chemical Corporation, Tokyo, Japan. S= stearate.
[b]DK Esters, produced by Dai-Ichi Kogyo Seiyaku Co. Ltd., Kyoto, Japan.
[c]HLB, hydrophilic-lipophilic balance.

Table III. Some Food Applications and Functions of Commercial Sugar Esters

Application	Function	Type[a]
Baked goods	soften crumbs, increase volume, inhibit staling and starch retrogradation, reduce stickiness to machine	S-1170 ~ S-1670; F-110 ~ 160
Margarine, butter, shortening	inhibit crystal growth, w/o emulsifier, prevent spattering	S-170~S-270;F-10~ F20W
Dairy products	o/w emulsifier, improve overrun, improve quality	all types
Confectionery	inhibit starch retrogradation, lubrication, prevent blooming, prevent adherence to the teeth	S-570~S-1670; F-20W~F-70
Processed meats and fish pastes	improve water holding capacity	all S-types; F-70~F-90
Canned drinks	antimicrobial agent, prevent flat-sour spoilage	P-1570~P1670[b]; F-14-~F-160

[a]See Table II for company sources.
[b]P = palmitate.

tenderness. Margarine and butter are w/o type food emulsion and will require a lipophilic SFE (low HLB value) for emulsification and stability. Antimicrobial properties of SFEs were excellently reviewed recently (28). Selected food applications of commercial sugar esters are illustrated in Table III.

Regulatory and Safety Aspects

Olestra is approved by the United States Food and Drug Administration (FDA) for use in savory snacks. Any future applications of olestra in other products will require separate petition and approval process. All other low calorie fats described here are not FDA approved. New molecules will also require testing for regulatory approval. They must be tested in both animals and humans. When appropriate, carcinogenicity, genotoxicity, reproductive and developmental toxicity tests must be performed. Both normal and at-risk individuals, such as people with compromised gastrointestinal tracts or other abnormalities, and children must be tested. Products containing olestra must be labeled.

The safety of olestra was established by evaluating its effects on the physiological function and the microflora of the gastrointestinal (GI) tract, on the absorption of drugs, and on the absorption and utilization of macro- and micronutrients in both animals and humans as reviewed recently (29). It has no effect on the absorption of flavonoids, polyphenols, and most phytochemicals (30). However, olestra may affect the absorption of vitamins A and E, and carotenoids. These effects can be offset by adding the vitamins to olestra-containing foods. Mild GI discomfort such as loose or soft stools, gas, or nausea, symptoms similar to those experienced with certain other foods or changed dietary habits may be expected with olestra consumption (29). Olestra effect on the environment was evaluated and found to be broken down by aerobic microorganisms and therefore is biodegradable in sewage sludge and soil environments (31).

Use of SFEs as food additives is regulated. SFE was approved in 1959 as food additives in Japan, 1974 in European Community, 1980 by FAO/WHO, and in 1983 by the FDA. Numerous studies indicate that SFE is safe and poses no significant problems.

Conclusion

Low calorie fats and structured lipids offer more choices to the consumer. More food applications of olestra and new product developments are expected in the future. Sugar esters have the potential for widespread use in food as emulsifiers or surfactants and as effective antimicrobial agents. The industry is, perhaps, not utilizing fully the potentials of sugar esters in food formulations.

Acknowledgments

Contributed by the Agricultural and Experiment Station, College of Agricultural and Environmental Sciences, The University of Georgia. Financial assistance was from Food Science Research and in part by an award from the International Life Sciences Institute.

Literature Cited

1.   AHA; Dietary guidelines for healthy adult Americans. *Am. Heart Assn. Circulation.* **1986,** *74,* 1465A.
2.   USDHHS; "The Surgeon General's Report on Nutrition and Health." *Publ. No. 88-50210.* U.S. Govt. Print. Office, Washington, D.C. 1988.

264

3.   AHA and USDHHS; Nutrition and your health: Dietary guidelines for Americans. 4th ed., Home and Garden Bulletin, No. 232. U.S. Dept. Agriculture and U.S. Dept. Health and Human Services, Washington, D.C. 1995.
4.   Mattson, F.H.; Nolen, G.A. *J. Nutr.* **1972,** *102,* 1171-1176.
5.   USDHHS; Food additives permitted for direct addition to food for human consumption: Olestra. Final rule. Food and Drug Administration, U.S., Dept. Health and Human Services, *Fed. Reg.* 1996, *61(20),* 3118-3173.
6.   Akoh, C.C. In *Carbohydrate Polyesters as Fat Substitutes;* Akoh, C.C.; Swanson, B.G., Eds.; Food Science and Technology Series 62; Marcel Dekker, Inc.: New York, NY, 1994, Vol. 1; pp 9-35.
7.   Rizzi, G.P.; Taylor, H.M. *J. Am. Oil Chem. Soc.* **1978,** *55,* 398-401.
8.   Shieh, C.J.; Koehler, P.E.; Akoh, C.C. *J. Food Sci.* **1996,** *61,* 97-100.
9.   Crouse, J.R.; Grundy, S.M. *Metabolism.* **1979,** *28,* 994-1000.
10.   Grundy, S.M.; Anastasia, J.V.; Kesaniemi, Y.A.; Abrams, J. *Am. J. Clin. Nutr.* **1986,** *44,* 620-629.
11.   Grossman, B.M.; Akoh, C.C.; Hobbs, J.K.; Martin, R.J. *Obesity Res.* **1994,** *2,* 271-278.
12.   Meyer, R.S.; Akoh, C.C.; Swanson, B.G.; Winter, D.B.; Root, J.M.; Campbell, M.L. *U.S. Patent* 4,973,489. 1990.
13.   Meyer, R.S.; Root, J.M.; Campbell, M.L.; Winter, D.B. *U.S. Patent* 4,840,825. 1989.
14.   Babayan, V.K. *Food Technol.* **1987,** *22,* 417-420.
15.   Bach, A.C.; Ingenbleek, Y.; Frey, A.; *J. Lipid Res.* **1996,** *37,* 708-726.
16.   Megremis, C.J. *Food Technol.* **1991,** *45,* 108-110, 114.
17.   Heird, W.C.; Grundy, S.M.; Hubbard, V.S. *Am. J. Clin Nutr.* **1986,** *43,* 320-324.
18.   Kennedy, J.P. *Food Technol.* **1991,** *76,* 78, 80, 83.
19.   Akoh, C.C. *INFORM.* **1995,** *6,* 1055-1061.
20.   Jensen, G.L.; Jensen, R.G. *J. Pediatr. Gastroenterol. Nutr.* **1992,** *15,* 382-394.
21.   Dehesh, K.; Jones, A.; Knutzon, D.S.; Voelker, T.A. *The Plant J.* **1996,** *9,* 167-172.
22.   Merolli, A.; Lindemann, J.; Del Vecchio, A.J. *INFORM.* **1997,** *8,* 597-602.
23.   CCC; Fat replacers: Food ingredients for healthy eating. Calorie Control Council, Atlanta, GA, 1996.
24.   Smith, R.E.; Finley, J.W.; Leveille, G.A. *J. Agric. Food Chem.* **1994,** *42,* 432-434.
25.   Kosmark, R. *Food Technol.* **1996,** *50,* 98-101.
26.   Artz, W.E.; Hansen, S.L. In *Carbohydrate Polyesters as Fat Substitutes;* Akoh, C.C.; Swanson, B.G., Eds.; Food Science and Technology Series 62; Marcel Dekker, Inc.: New York, NY, 1994, Vol. 1; pp 197-236.
27.   Nakamura, S. *INFORM.* **1997,** *8,* 866-874.
28.   Marshall, D.L.; Bullerman, L.B. In *Carbohydrate Polyesters as Fat Substitutes;* Akoh, C.C.; Swanson, B.G., Eds.; Food Science and Technology Series 62; Marcel Dekker, Inc.: New York, NY, 1994, Vol. 1; pp 149-167.
29.   Lawson, K.D.; Middleton, S.J.; Hassall, C.D. *Drug Metabolism Rev.* **1997,** *29,* 651-703.
30.   Cooper, D.A.; Webb, D.R.; Peters, J.C. *J. Nutr.* **1997,** *127,* 1699S-1709S.
31.   Allgood, G.S.; Shimp, R.J.; Federle, T.W.; Annual Meeting of the Society of Environmental Toxicology and Chemistry, Denver, Co, 1994, Oct. 30-Nov. 4.

Chapter 17

Intestinal Absorption and Physiologically Functional Food Substances

M. Shimizu[1] and K. Hashimoto[2]

[1]Department of Applied Biological Chemistry, Graduate School of Agricultural and
Life Sciences, The University of Tokyo, Bunkyo-ku, Tokyo 113, Japan
[2]Department of Bioproductive Sciences, Utsunomiya University, Mine-cho,
Utsunomiya 321, Japan

Intestinal absorption is an important process which determines the
bioavailability of the nutrients and physiologically active substances in
food. Although the existence of various pathways for intestinal
absorption has been already reported, the effect of food-derived
substances on the intestinal absorption mechanism is poorly understood.
A monolayer of human intestinal epithelial Caco-2 cells cultured on a
semipermeable filter was used as a model of the intestinal epithelium in
an attempt to detect those food substances which would affect the
intestinal epithelial permeability. Among the various food samples
investigated, an extract of sweet pepper (*Capsicum annuum L.var.
grossum*) was found to contain certain substances which could modulate
the paracellular permeability. The permeability for water-soluble
substances of low molecular weight, including oligopeptides, was
facilitated by treating the cell monolayer with this extract. Molecular
and functional properties of the active substances in this extract were
also investigated.

A variety of bioactive food substances which can exert a physiological effect on the
endocrine, nerve, immune and circulation systems has been discovered during these last
two decades. Some of them are expected to be used as ingredients for physiologically
functional foods (*1*). However, the effect of food substances on the intestinal functions,
particularly on the absorption and transport functions, has not been well studied,
although intestinal absorption is an important process that determines the bioavailability
of nutrients and bioactive substances derived from food. Some food substances present
or produced in the intestinal tract could affect intestinal functions, thereby modulating
the intestinal absorption of nutrients or other components; examples of such substances
are presented in this contribution.

Intestinal epithelium and absorption mechanism
Nutrients are, in most cases, absorbed at the small intestinal epithelium which consists of several types of cell, the majority being absorptive columnar cells (2). The absorptive cells form a monolayer on the epithelial surface of intestinal villi. Nutrients present in the intestinal tract are absorbed from the apical (mucosal) side of this epithelial cell monolayer and transported to the basolateral (serosal) side (2). This absorption is mainly accompanied via the following four pathways (Figure. 1): (1) Transporter-mediated transport: there are several transporter proteins in the apical/basolateral cell membranes (3) which selectively transport such nutrients as glucose, amino acids and di(tri)peptides. (2) Transcytosis: relatively high molcular-weight components such as proteins may be apically taken into a cell by endocytosis, intracellularly transported via the transcytotic vesicles, and released to the basolateral side of the cell by exocytosis (4). (3) Itracellular passive diffusion: such hydrophobic nutrients as fat-soluble vitamins are diffused into the cells across the hydrophobic apical cell membrane, intracellularly transported via certain binding proteins, and released to the basolateral side by diffusion (5). (4) Paracellular passive diffusion: water-soluble substances with relatively low molecular weights, including oligopeptides (6) and such mineral ions as Ca^{2+} (7), may be transported through the junction between the cells by diffusion. Under certain conditions, the paracellular pathway is thought to be responsible for the transepithelial transport of such nutrients as glucose and amino acids as well (8).

Paracellular pathway and tight junction
The intestinal epithelial cell layer functions as a gate for nutrients, as well as a barrier which separates the luminal (apical) and vascular (basolateral) fluid compartments. The barrier function is substantially dependent on the intercellular tight junction (TJ) that restricts the flow through the paracellular pathway. On the other hand, it is becoming clear that the paracellular pathway also provides a highly dynamic transport route for certain ions and hydrophilic molecules, thus contributing to the intestinal absorption of various nutrients (9). As already mentioned, various water-soluble substances are thought to be transported via this route. There are several physiologically active oligopeptides derived from food proteins, such as the opiatic, immunostimulative, and hypotensive peptides from milk proteins (10). The apical-to-basolateral transport of these oligopeptides in an intact form is mainly via the paracellular pathway (6,11,12), which must therefore be important for the physiologically active peptides to express their activity *in vivo*.

The TJ structure had not been clear until several TJ-related proteins were cloned, sequenced and characterized during the past decade. The protein which directly participates in the cell-cell junction was first cloned by Furuse *et al.* (13) and designated as occludin. Occludin is thought to be associated directly or indirectly with several intracellular proteins such as ZO-1, ZO-2, 7H6 and cingulin (14). These TJ-related intracellular proteins are further associated with cytoskeletal structures such as actin filaments (Figure 1, inset). These findings, indicating the presence of a complicated network of TJ-related proteins, suggest that TJ could be physiologically regulated in its permeability by intracellular messengers or external factors including food-derived substances.

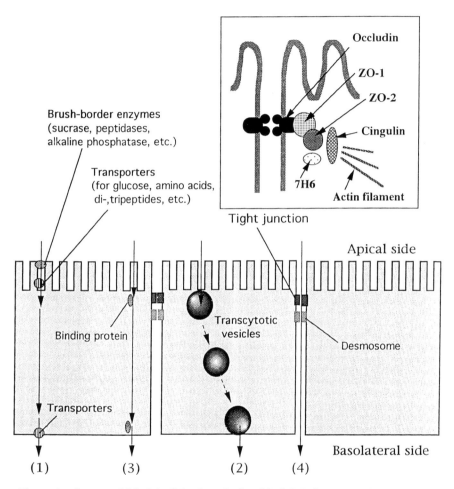

Figure 1. Structural Model of the Intestinal Epithelial Cell Layer and of the Four Major Pathways (1-4) for Nutrient Absorption.
The inset shows a model for the organization of the tight junction.

We were interested in the effect of food substances on the TJ permeability, because those substances might facilitate or suppress the intestinal absorption of nutrients and functional food components. TJ-regulatory substances from various food materials were therefore screened.

Caco-2 human intestinal epithelial cell line

In order to evaluate the effect of food substances on TJ permeability, a good assay system is essential. Since an experiment that uses an excised intestinal tract or an *in vivo* experiment may not be suitable for the screening test, we used a cell culture system which mimicked the intestinal epithelium. The human colon adenocarcinoma cell line, Caco-2 (*15*), has been observed to have spontaneous differentiation ability and to exhibit various enterocytic characteristics; e.g., expressing brush-border enzymes such as sucrase (*16*), nutrient transporters such as the PepT1 peptide transporter (*17*), and intercellular TJ (*18*). Culturing this cell on a semipermeable filter forms an epithelium-like monolayer separating the apical and basolateral compartments (Figure 2).

The TJ permeability was evaluated by measuring the transepithelial electrical resistance (TEER) (Figure 2). The resistance of the paracellular route is much lower than that of the transcellular one, so that the TEER value across the epithelial monolayer mainly reflects the resistance offered by TJ. As shown in Figure 3, the paracellular transport of a fluorescent dye marker, Lucifer Yellow (MW 457), in the Caco-2 cell monolayers declined substantially as the TEER value increased (*19*). This suggests that the TEER value would be a good marker to evaluate the TJ permeability.

Screening of TJ-regulatory food substances

Vegetable, fruit, bean, seaweed and other food material samples were homogenized in 4 volumes of deionized water at room temperature. Each homogenate was centrifuged and the supernatants then freeze-dried. Caco-2 cells were monolayer-cultured on a semipermeable filter (Millicell-HA; Millipore). After 9 days of culture, each sample was added to the apical side of the cell layer (Figure. 2) to give a final concentration of 5 mg/ml in a Ringer solution which had been adjusted to pH 5.5. TEER of the monolayer was monitored with Millicell-ERS equipment (Millipore) after a 30-min period of incubation (*19*). Among more than 80 food samples, the extracts of sweet pepper, horseradish, ginger, garlic and some mushrooms (fungi) significantly decreased the TEER value. The cell viability determined by the trypan blue exclusion and lactate dehydrogenase release methods was, however, significantly reduced after the cells had been treated with the extracts of ginger, garlic and some of the mushrooms including shimeji (*Lyophyllum aggregatum*) and nameko (*Pholiota nameko*), indicating that the TEER decrease caused by these samples was due to their cytotoxic properties. On the other hand, the extracts of sweet pepper, horse-radish and enokitake (winter mushroom; *Collybia velutipes*) decreased the TEER value without exerting any cytotoxic effect on the cells. The active substance contained in sweet pepper (*Capsicum annuum L.var. grossum*) was then isolated and characterized.

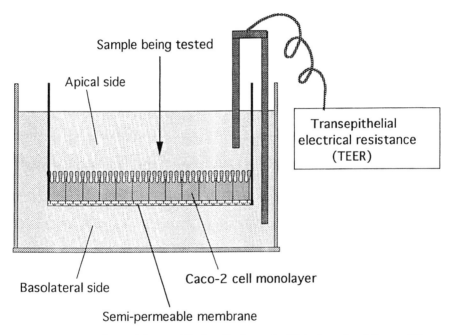

Figure 2. Experimental System Used to Evaluate the Tight Junctional Permeability.

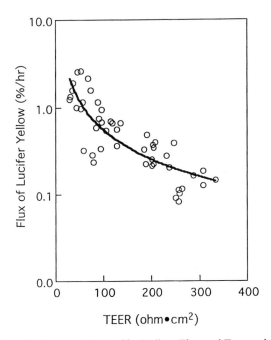

Figure 3. Relationship between the Lucifer Yellow Flux and Transepithelial Electrical Resistance (TEER) in a Caco-2 Cell Monolayer.

Caco-2 cell monolayers with different TEER values were prepared by culturing the cells under different conditions. After the TEER value of each had been determined, a Lucifer Yellow solution was added to the apical compartment (see Figure. 2). An aliquot of the medium was collected from the basolateral compartment after 1 h of incubation, and Lucifer Yellow was determined by measuring its fluorescence intensity (*19*).

Active fraction from the sweet pepper extract

The crude sweet pepper extract caused a dose-dependent reduction in TEER of the Caco-2 cell monolayer. The TEER-decreasing activity of the sweet pepper extract remained unchanged after heating the extract at 100°C for 5 min. Digestion of the extracted material with Pronase did not affect the activity, suggesting that the active substance in sweet pepper was not a proteinaceous material. In order to isolate the active substance, the extract was first applied to a DEAE-ion exchange column. The passed-through fraction having TEER-decreasing activity was then applied to a Sep-Pak-C18 cartridge (Millipore), and the adsorbed materials were recovered by stepwise elution with actonitrile. The highest activity was detected in the Sep-Pak fraction eluted with between 30% and 50% acetonitrile in water (20).

The active fraction was further fractionated by Superose 12 gel chromatography (Pharmacia), and the TEER-decreasing activity of the fractions was measured (20). Figure 4 shows the elution pattern monitored at 280 nm and the TEER value of the Caco-2 monolayer after being treated with each fraction. The fraction from No. 25 to No. 29 (named the SP fraction) decreased the TEER value to less than 70% of the value at zero time. The UV absorption of the SP fraction at 280 nm was very weak, but a considerable amount of carbohydrate was detected in this fraction, suggesting that the active substance was a thermally stable complex carbohydrate or related substance. The yield of the SP fraction was 27.5±2.8 mg from 100 g (wet weight) of sweet pepper (20).

The effect of the SP fraction on the permeability of the cell monolayer was examined by using a leakage marker (Lucifer Yellow), oligopeptide (bradykinin) and protein (horseradish peroxidase). In the cell monolayer treated with the SP fraction (200µg/ml) for 1 hr, the permeability for Lucifer Yellow (MW 457) and bradykinin (MW 1060) was about 2 times higher than that in the control monolayer (Figure 5). In contrast, the permeability for horseradish peroxidase (MW 40,000) remained unchanged. This selective increase in the permeability suggests that opening of the paracellular route was not due to simple TJ destruction, but was instead due to a subtle modulation of the TJ structure.

Figure. 6 shows electron microscopic images of TJ in a Caco-2 cell monolayer before and after treating the cell layer with the SP fraction. Treating the cell monolayer with the SP fraction widened the junctional space (Figure 6B, arrow), while the apical portion of TJ between two adjacent cells was still attached to both (Figure 6B, arrowhead). This observation also indicates that the increased permeability caused by the SP fraction was not due to any damage to the cell monolayer, but was rather based on a cellular mechanism that regulated the TJ structure.

The effect of the SP fraction was observed not only in the Caco-2 cell monolayer system but also in an experiment with rat small intestinal mucosa (unpublished results). TEER of the intestinal mucosa was decreased by incubating with the SP fraction in a dose-dependent manner, although the sensitivity of the intestinal mucosa to the SP fraction was lower than that of the Caco-2 monolayer. This decrease in TEER was promptly eliminated by washing out the SP fraction from the mucosa, the TEER value being restored to the initial level.

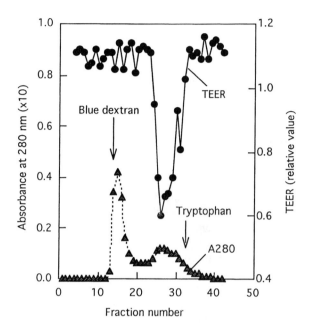

Figure 4. Purification of the TEER-decreasing Substances by Superose 12 Chromatography.
The isolated fractions were each assayed for their TEER-decreasing activity and protein content.

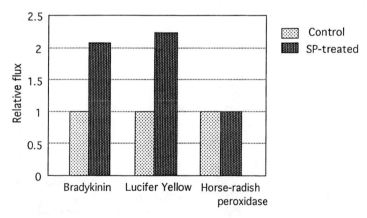

Figure 5. Effect of the SP fraction on the Permeability of the Caco-2 Cell Monolayer.
The apical-to-basolateral flux was measured by using Caco-2 cell monolayers that had been pretreated with the SP fraction for 1 h, the results being presented as relative to the values in the control cell layer.

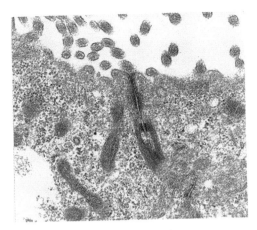

(A) Control

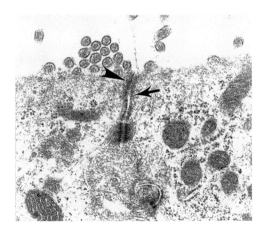

(B) Treated with SP-fraction

Figure 6. Electron Microscopic Images of the Tight Junction in Caco-2 Cell Monolyers.
(A) Control cell layer; (B) cell layer preincubated with the SP fraction (500 µg/ml for 10 min). The junctional space has been opened by treating with the SP fraction (B; shown by an arrow). The arrow head indicates the site of apparent fusion of the adjacent cell membranes.

Isolation and characterization of the active substance in the SP fraction

The active substance in the SP fraction was further purified by reverse-phase HPLC with an ODS column and analyzed by HPLC (20). The analytical results indicated at least five major components (SP-1-5). SP-1-5 corresponding to the five major peaks by HPLC were purified by rechromatography, and the structure of SP-1-5 was then analyzed by ^{1}H- and ^{13}C-NMR and FAB-MS. SP-1-5 had a similar structure and were identified as capsianosides, diterpene esters of monomeric diterpene glycosides (21). SP-1, 2, 3, 4 and 5 corresponded to capsianoside C, A(D), E, B and F, respectively. The TEER-decreasing activity varied among these capsianosides, the order of activity being F>B>E>A(D)>C. The structure of capsianoside F, the most potent one of the TJ-modulatory substances in sweet pepper, is shown in Figure 7.

It is well known that Ca^{2+} removal affects the Ca^{2+}-dependent adhesion molecules, thereby inducing the disassembly of TJ components(22). Capsianoside F was observed to have some Ca^{2+}-chelating activity, being about one-tenth that of EDTA. However, its TEER-decreasing activity was not altered by adding excess Ca^{2+} to the medium, suggesting that the increased TJ permeability by capsianosides was not due to their Ca^{2+}-chelating activity (20).

Since the TJ-related protein complex, involving ZO-1, ZO-2 and occludin, associates with cytoskeletal elements such as actin filaments, any perturbation of the cytoskeletal structure would induce a change in TJ permeability (14). We have observed that a capsianoside F treatment (50 μg/ml for 30min) decreased the intracellular G-actin content to 60% of the original level (20). In contrast, the relative F-actin content was increased by 16% by a Capsianoside F treatment for 15 min. A morphological change in the actin ring was also observed. These findings strongly suggest that capsianosides stimulated the reorganization of the actin filaments, by which the TJ permeability was increased (Figure 8). The participation of protein kinase C in this reorganization of the actin filaments has also been suggested (20).

Other food-derived substances affecting TJ permeability

We have found that enokitake (winter mushroom) also contained a substance that increased TJ permeability (unpublished results). The active substance in enokitake was identified to be a protein with an approximate molecular weight of 30,000. Unlike capsinosides, this protein was heat-labile, the activity being eliminated by heating at 100°C for 3 min. The mechanism for the increase in TJ permeability by this protein is now under investigation, but it seems to be different from that of capsianosides. Some of the fatty acids and their glycerides, particularly those with a medium-chain fatty acid such as caprylic acid, have also been reported to increase the paracellular permeability (23).

Concluding remarks

Capsianosides would be useful for enhancing the permeability for nutrients or other biologically important hydrophilic substances across the intestinal mucosa. As capsianosides maintain their activity in the presence of a high concentration of Ca^{2+}, they would, for example, be able to enhance Ca^{2+} permeability.

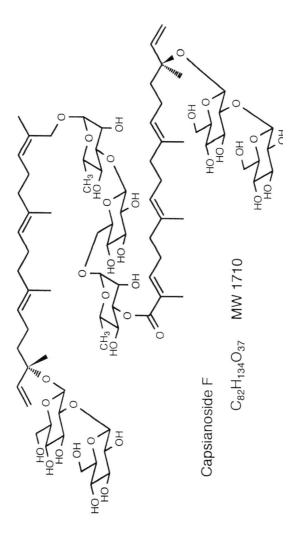

Capsianoside F

$C_{82}H_{134}O_{37}$ MW 1710

Figure 7. Structure of Capsianoside F.

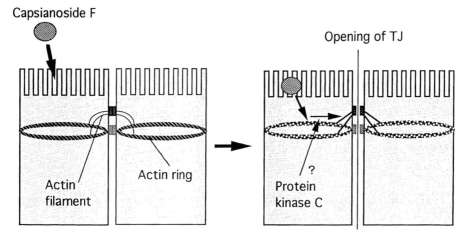

Figure 8. Schematic Model Showing How Capsianoside F Affects the Permeability of the Intestinal Epithelial Cell Layer by Modulating the Cytoskeletal Structure.

Like sweet pepper and enokitake, materials used as food ingredients may contain a variety of substances which could modulate intestinal absorption. In addition to the TJ-modulatory substances just described, there could be some substances which affect the transporter function. Those which inhibit glucose transporters in the intestinal epithelium, such as gymnemic acid (*24*), may be useful to regulate glucose absorption in diabetic patients. Inhibitory substances against intestinal digestive enzymes, such as α-glucosidases, would also affect the intestinal absorption of carbohydrate, thereby suppressing an excessive calorie intake (*25*). Hydrophobic peptides isolated from soybean proteins are known to inhibit cholesterol absorption in the intestinal tract (*26*).

Studies on food-derived substances which affect the intestinal absorption would provide valuable information for the design and development of new functional foods.

Literature Cited
1. Arai, S. *Biosci. Biotech. Biochem.* **1996**, *60*, 9-15.
2. Cheng, H.; Leblond C. P. *Am. J. Anat.* **1974**, *141*, 461-480, 537-562.
3. Hediger, M. A. *J. Exp. Biol.* **1994**, *196*, 15-49.
4. Heyman, M.; Desjeux, J. F. *J. Pediat. Gastroenetrol. Nutr.* **1992**, *15*, 48-57.
5. Ong, D. E. *J. Biol. Chem.* **1984**, *259*, 1476-1482.
6. Pappenheimer, J. R.; Dahl, C. E.; Karnovsky, M. L.; Maggio, J. E. *Proc. Natl. Acad. Sci. USA* **1994**, *91*, 1942-1945.
7. Karbach, U. *J. Nutr.* **1992**, *122*, 672-677.
8. Atisook, K.; Madara, J. L. *Gastroenterology* **1991**, *100*, 719-724.
9. Schneeberger, E. E.; Lynch, R. D. *Am. J. Physiol.* **1992**, *262*, L647-L661.
10. Fiat, A-M.; Jolles, P. *Mol. Cell. Biochem.* **1989**, *87*, 5-30.
11. Adson, A.; Raub, T. J.; Burton, P. S.; Barsuhn, C. L.; Hilgers, A. L.; Audus, K. L.; Ho, N. F. H. *J. Pharm. Sci.* **1994**, *83*, 1529-1536.
12. Shimizu, M.; Tsunogai, M.; Arai, S. *Peptides* **1997**, *18*, 681-687.
13. Fruse, M.; Hirase, T.; Itoh, M.; Nagafuchi, A.; Yonemura, S.; Tsukita, S.; Tsukita, S. *J. Cell Biol.* **1993**, *123*, 1777-1788.
14. Citi, S. In *Molecular Mechanisms of Epithelial Cell Junctions,* Citi, S. Ed.; R.G. Landes Co., Austin, 1994; pp. 83-106.
15. Fogh, J.; Fogh, J. M.; Orfeo, T. *J. Natl.Cancer Inst.* **1977**, *59*, 221-226.
16. Hashimoto, K.; Shimizu, M. *Cytotechnology* **1993**, *13*, 175-184.
17. Liang, R.; Fei, Y-J.; Prasad,P. D.; Rammamoorthy, S.; Han, H.; Yang-Feng, T.; Hediger, M. A.; Ganapathy, V.; Leibach, F. H. *J. Biol. Chem.* **1995**, *270*, 6456-6463.
18. Hidalgo, I. J.; Raub, T. J. and Borchardt, R. T. *Gastroenterology* **1989**, *96*, 736-749.
19. Hashimoto, K.; Matsunaga, N.; Shimizu, M. *Biosci. Biotech. Biochem.* **1994**, *58*, 1345-1346.
20. Hashimoto, K.; Kawagishi, H.; Nakayama, T.; Shimizu, M. *Biochim. Biophys. Acta* **1997**, *1323*, 281-290.
21. Izumitani, Y.; Yahara, S.; Nohara, T. *Chem. Pharm. Bull.* **1990**, *38*, 1299-1307.
22. Jones, K. H.; Senft, J. A. *J. Histochem. Cytochem.* **1985**, *33*, 77-79.

23. Shima, M.; Todou, K.; Yamaguchi, M.; Kimura, Y.; Adachi, S.; Matsuno, R. *Biosci. Biotech. Biochem.* **1997**, *61*, 1150-1155.
24. Yoshioka, S. *J. Yonago Med. Ass.* **1986**, *37*, 142-154.
25. Honda, M.; Hara, Y. *Biosci. Biotech. Biochem.* **1993**, *57*, 123-124.
26. Sugano, M.; Goto, S.; Yamada, Y.; Yoshida, K.; Hashimoto, Y.; Matsuno, T.; Kimoto, M. *J. Nutr.* **1990**, *120*, 977-985.

INDEXES

Author Index

Subject Index

A

Ab initio, computer modeling type, 39

Acylation, chemical modification method, 100, 102

Alkyl glycosides fatty acid polyester, possible replacement for fats and oils, 257

Amaranth grain
amino acid, lysine, content, 67
commercialization of, 67
food source by old cultures, 66
protein content of, 66–67
three species of, 66
See also High-protein amaranth flours

Amino acids
composition of flaxseed protein isolate, 103, 106*t*
content in high-protein amaranth flours, 76*t*
partial specific θ values, 11*t*
protein composition and relationship to Θ, 10–11

Animal proteins. *See* Goat meat proteins

Aspartic proteinases. *See* Milk-clotting enzymes

B

Bacillus stearothermophilus neutral protease. *See* Computer-aided optimization of site-directed mutagenesis

Beef protein
comparison to goat, 222–225
See also Goat meat proteins

Benefat, brand name of structured lipid, salatrim, 259

Biological catalyst design. *See* Computer-aided optimization of site-directed mutagenesis

Bovine serum albumin (BSA)
active in decreasing interface free energy, 6
adsorption at interfaces, 3–4, 5*f*
surface concentration versus surface pressure plot, 8*f*

C

Caprenin, (caprocaprylobehenin) structured lipid, 259

Capsianosides

model of effect on permeability of intestinal epithelial cell layer, 276*f*
modulating intestinal absorption, 274, 277
potent tight-junction modulating substances in sweet pepper, 274, 275*f*
See also Intestinal absorption

β-Casein
adsorption at interfaces, 3–4, 5*f*
hydrophobic and random-coil-like protein, 6
surface concentration versus surface pressure plot, 8*f*
tendency for rapid conformation changes, 6–7

Catalyst design. *See* Computer-aided optimization of site-directed mutagenesis

Cholesterol-reduced egg yolk low-density lipoprotein (CR-LDL)
analysis of phospholipids and cholesterol at interface, 207
changes in emulsion mean particle size as function of storage time, pH, and NaCl concentration, 215*f*
cholesterol extraction with β-cyclodextrin, 206
cholesterol important in stabilization of LDL emulsions, 216
cholesterol reduction from egg yolk LDL, 207–208
composition of cholesterol and phospholipids at oil-in-water interface as function of LDL concentration and pH, 213*f*
effect of β-cyclodextrin (CD):cholesterol molar ratio on reduction of cholesterol from egg yolk LDL, 208*t*
emulsifying properties determination method, 206–207
emulsifying properties of CR-LDL, 208–211
particle size as function of LDL concentration and pH, 209*f*
phospholipids and cholesterol composition at interface, 211, 214
preparation method, 206
protein composition at oil-in-water interface, 211, 212*f*
stability of LDL and CR-LDL emulsions, 214, 216
surface protein coverage of emulsions as function of LDL concentration, 210*f*
technologies for LDL reduction, 205–206

Color, acylated flax protein isolates, 103, 107*t*, 108

Compressibility, descriptor of protein's surface activity, 9–10

Highlights from ACS Books

Desk Reference of Functional Polymers: Syntheses and Applications
Reza Arshady, Editor
832 pages, clothbound, ISBN 0–8412–3469–8

Chemical Engineering for Chemists
Richard G. Griskey
352 pages, clothbound, ISBN 0–8412–2215–0

Controlled Drug Delivery: Challenges and Strategies
Kinam Park, Editor
720 pages, clothbound, ISBN 0–8412–3470–1

Chemistry Today and Tomorrow: The Central, Useful, and Creative Science
Ronald Breslow
144 pages, paperbound, ISBN 0–8412–3460–4

Eilhard Mitscherlich: Prince of Prussian Chemistry
Hans-Werner Schutt
Co-published with the Chemical Heritage Foundation
256 pages, clothbound, ISBN 0–8412–3345–4

Chiral Separations: Applications and Technology
Satinder Ahuja, Editor
368 pages, clothbound, ISBN 0–8412–3407–8

Molecular Diversity and Combinatorial Chemistry: Libraries and Drug Discovery
Irwin M. Chaiken and Kim D. Janda, Editors
336 pages, clothbound, ISBN 0–8412–3450–7

A Lifetime of Synergy with Theory and Experiment
Andrew Streitwieser, Jr.
320 pages, clothbound, ISBN 0–8412–1836–6

Chemical Research Faculties, An International Directory
1,300 pages, clothbound, ISBN 0–8412–3301–2

For further information contact:
Order Department
Oxford University Press
2001 Evans Road
Cary, NC 27513
Phone: 1-800-445-9714 or 919-677-0977
Fax: 919-677-1303